TRANSACTIONS

of the

American Philosophical Society

Held at Philadelphia for Promoting Useful Knowledge

VOLUME 79, Part 3

Towards a Rational Historiography

LIONEL GOSSMAN

Department of Romance Languages, Princeton University

THE AMERICAN PHILOSOPHICAL SOCIETY

Independence Square, Philadelphia

1989

Library of Congress Catalog
Card Number 89-84933
International Standard Book Number 0–87169–000–0
US ISSN 0065–9746

Foreword

Since it reached the state in which it appears here in the summer of 1986, I have had the benefit of comments on my essay from a number of colleagues and friends, notably from the members of a faculty group studying the relations between literature and history at the Claremont Colleges in California. At the end of the three-hour seminar with the Claremont group in fall, 1987, it became clear to me not only that there are a number of unresolved issues in the essay, but that these issues are also at the center of other work on which I am currently engaged and which had seemed to me, until then, unconnected with the essay on historiography.

Underlying the argument that historiography cannot be subsumed under a poetics or a rhetoric (in the sense of a system of purely linguistic or literary tropes) is a larger claim, namely that a wide range of activities, extending from literary criticism (my own professional specialization), through legal debate, theology, ethics, politics, psychology, and medicine to the natural sciences, all constitute rational practices, even if there is considerable variation in the degree of formalism and rigor and in the type of argument most commonly employed in each of these different fields of inquiry.[1] That is why in the essay I emphasize the practice or process of doing history rather than the product and, admittedly with some provocativeness, take Bayle rather than Gibbon or Michelet as my model historian.

The position I try to defend owes much to several works mentioned in the essay: Hans Albert's *Treatise on Critical Reason*, the writings of the Belgian philosopher Chaim Perelman, and Stephen Toulmin's *Human Understanding*. Since completing the essay I have come across a remarkably well constructed manual, entitled *An Introduction to Reasoning*, which Toulmin and two associates, Richard Rieke and Allan Janik, prepared for undergraduate teaching. My own views are formulated in that work with such clarity and simplicity that I cannot do better than quote from it here.

The essential locus of reasoning is seen to be a *public, interpersonal,* or *social* one. Wherever an idea or thought may come from, it can be examined or criticized "rationally"—by the standards of "reason"—only if it is put into a position where it is open to public, collective criticism. Reasoning is thus not a way of *arriving*

---

[1] For an admirable account of criticism as a rational practice, see the chapter entitled "Literary Aesthetics and Literary Practice" in Stein Haugom Olsen, *The End of Literary Theory* (Cambridge: Cambridge Univ. Press, 1987), especially 7–16.

at ideas but rather a way of *testing ideas critically.* . . . It is a collective and continuing human transaction in which we present ideas or claims to particular sets of people within particular situations or contexts and offer the appropriate kinds of "reasons" in their support. So reasoning involves dealing with claims with an eye to their contexts, to competing claims, and to the people who hold them. It calls for the critical evaluation of these ideas by shared standards; a readiness to modify claims in response to criticism; and a continuing critical scrutiny both of the claims provisionally accepted and of any new ones that may be put forward subsequently. A "reasoned" judgment is thus a judgment in defense of which adequate and appropriate reasons can be produced.[2]

The authors of an *Introduction to Reasoning* go on to discuss differences in the degree of formality required by different activities (extremely rigorous in the law, moderate in the sciences, rather loose in literary or artistic interpretation), in the precision and exactitude of arguments, in modes of resolution. They are at pains to emphasize that exactitude of the type associated, since the Greeks, with mathematics is not a condition of "rationality." This kind of exactitude, they point out, may find a place at some point in *any* field or enterprise. "Even in a seemingly 'informal' activity, such as literary criticism, certain procedures may be conveniently computerized." On the other hand,

no enterprise can rely on this kind of strict or exact argumentation *alone*. Even the exact sciences such as physics involve phases of unformalizable interpretation that depend in part on the exercise of personal judgment. So it is a mistake to imagine that argumentation is always "formal" in some fields (e.g., natural science) and always "informal" in others (e.g., aesthetics).[3]

What appeals to me in the idea of reason as a practice is its open, liberal, and democratic character. Historiography as a rational practice supposes a community of participants rather than the *anomie* of a world in which every man is his own historian or, at best, the relation of hero and follower that appears to be implied by privileging the historical *text*. Instead of aggravating the tension between reason and decision, what is sharable or common and what is "creative" and "original," philosophers like Toulmin and Albert aim to soften it by arguing that decisions, while still having the character of decisions, may be more or less reasonable. Toulmin emphasizes in particular that in most human activities, people are called on to make decisions, for which they feel and will be held accountable, without being in possession of all the relevant information and without any hope of arriving at the kind of certainty sought by Descartes, for example. (It is most probably not incidental that Descartes chose a life of exile and anonymity, almost entirely devoted to theoretical reflection, and with as little participation as possible in public affairs, where uncertainty remained in his view most intractable. Equally, it is

---

[2] Stephen Toulmin, Richard Rieke, Allan Janik, *An Introduction to Reasoning* (New York: Macmillan, 1979), 9.
[3] *Introduction to Reasoning*, 197.

worth recalling Pascal's argument, against Descartes, that most of the issues on which human beings have to act do not admit of certainty, and that reasonable behavior therefore consists in reaching as considered a decision as possible, not in deferring action until absolute certainty has been attained.[4]) Similarly, instead of presenting different "paradigms" as uncommunicating, Toulmin tries to show that one develops out of criticism of another and that the scientific community is often in fact quite quick to recognize and adopt a "better" or more plausible hypothesis. Finally, instead of concluding from the absence of absolute foundations for our knowledge (both the traditional theories of truth—the correspondence theory and the coherency theory—having been found inadequate) that total epistemological relativism and skepticism are the only possible attitude, Toulmin develops a view of knowledge and research as moderately orderly and regulated activities that build by collectively achieved and approved methods on collectively achieved and shared beliefs or "truths" through the refinement, criticism, modification, and, from time to time, rapid transformation of these truths and methods.[5] What we have at any given point is not absolutely certain knowledge, but the only knowledge we can have, the knowledge produced by our scientific culture. That knowledge, however, is neither arbitrary, nor whimsical, nor valueless. As long as we wish to participate in a human community and contribute to it by having our views considered by it and possibly adopted by it, we cannot just take up any position we like. To be sure, we are always at liberty to do that, but then we should not expect to have our position reviewed for possible adoption by others.

There may, for example, be an ideological component in an historical argument, or even in an argument, such as I try to develop in the essay, *about* historical argument. Rationality consists, however, not in removing the ideological component altogether or in withdrawing from all areas where ideology intervenes, but in being willing to have an ideological stand scrutinized and criticized and, eventually, to modify it. Like decisions, values and ideologies cannot be rationalized *away*; but they may be more or less rational, more or less open to criticism and discussion, more or less plausible and defensible.

As a citizen of a democratic community, a literary scholar and historian, and a teacher, I believe we can and should be accountable in history and in literary criticism, as in anything else. I personally consider that one of my tasks as a teacher is to get students to see that I have no authority other than the plausibility and pertinence of my arguments, that they likewise must be prepared to argue for any view they put

---

[4]See in particular *Pensées* (Brunschwieg ed.), #233 (the wager) and #234. "Combien de choses on fait pour l'incertain, les voyages sur mer, les batailles. Je dis donc qu'il ne faudrait rien faire du tout, car rien n'est certain [. . .] Or, quand on travaille pour demain, et pour l'incertain on agit avec raison; car on doit travailler pour l'incertain . . ." (#234)
[5]*Introduction to Reasoning*, 129–35.

forward, and that in general, even if it is impossible to provide an absolutely watertight demonstration of a position, we do and should provide reasons for maintaining it. I think that our practice as professionals does in fact correspond to this model. I also think we ought to strive to make it correspond to it more and more.

In the early eighteenth century the French writer Marivaux criticized what he called the classical "Auteur" as some one who simply pronounces what he has to say without any consideration of his audience, and promised, in typical Enlightenment manner, that he would proceed differently, "offering" his writings to the public for its consideration, "proposing" ideas to his readers that they were to be free to accept, criticize, or reject. Marivaux's intention, in other words, was to occupy the same plane as his readers and to renounce all charisma, all desire to overpower, overwhelm or dominate them. Literature was to be demythified or desacralized along with everything else. The writer's claim to inhabit the atemporal sphere of the "classical"—the Olympian summit, towering above any particular audience, any particular response, any particular context—was to be abandoned for a practice of immediate engagement and exchange with a particular audience and for a less exalted, more down-to-earth view of literature as a social institution.

Marivaux's plans for literature itself correspond strikingly to the model of literary criticism and history that I have outlined in my essay. They also raise for me, as a teacher of literature, the question of the status of literary discourse and the literary text. I may well be able to argue plausibly for a rational literary criticism (according to the modest definition of rationality I have adopted), as well as for a rational historiography, but where does that leave literature itself, including those privileged historiographical texts that continue to function as literature after they have ceased to contribute actively to ongoing historical research? In other words, even if I can make a case for considering literary criticism and history as rational practices, whose hypotheses and results are subject to constant criticism and review, can literature itself be included in the category of rational practices, as Marivaux appears to propose, or is it somewhere above and beyond reasoning, above and beyond history itself? Are literary works perhaps irretrievably "charismatic"?

I am too much of a historian not to want immediately to question the concept of "literature" as used here. The idea that "literature" is an essence, an ahistorical category, is one I instinctively resist. It seems highly likely that it owes something to the capacity of written texts—as distinct, say, from oral or folklore productions—to survive their historical contexts and audiences unchanged and to continue to be available, like works of art in the privileged space of the museum, to successive generations of readers. I also happen to think it is not unrelated to the profound disillusionment that affected many European writers, artists, and intellectuals around the time of the 1848 Revolutions.

The collapse of historical optimism—I mean of the conviction that history is a continuity with a *sens* (in the French sense of both meaning and direction), of which moments of rupture and revolution are themselves part; that, correspondingly, "nations" or "peoples," embracing all social classes, and "humanity," embracing all the peoples and nations of the earth, are historical realities in the process of becoming; and finally, that nature, the life of the individual person, and the historical and social world, despite the different specialized sciences used to study each of them, constitute a unity—appears to have entailed, for literature, a divorce between rhetoric and poetic, the language of mundane communication and the poetic use of language. In France, Hugo and Michelet continued to address their audiences as citizens and to write about matters of public concern in literary and poetic language, but in the 1850s and 1860s they were already relics of another age. The modern idea of literature arose, it would seem, in the aftermath of 1848, out of a determination to salvage language in a social and historical context, in which grandiloquent public talk about national unity, fraternity, progress, and humanity was perceived as a hypocritical veil thrown over the crassest cupidity and class conflict, at best a vulgar and culpable illusion. To engage in public discourse in such circumstances was to subject language to the corruption and prostitution that had come to be seen as inseparable from public life. The desire to protect literature from commercialization, "commodification," and the encroachments of an ever expanding market economy, that is to say from "history," was almost certainly another—closely connected—factor contributing to our modern idea of literature. What I am suggesting is that literature may not always have been something that refuses to "speak" to us in the language of exchange and argument characteristic of all the activities Toulmin and his associates define as rational. There may have been a time when the boundary between "literature" and other forms of discourse was more porous. But even if literature has *become* that which it is widely understood to be, one is bound to ask what it means *now* to be a literary scholar or critic or interpreter, that is, some one who translates the language of the literary work and makes it accessible, who, by "interpreting" and "explaining" it, brings it down into the dusty marketplace of public discourse and exchange of ideas, of "culture," and talk, and culture-talk, from which it sought to extricate itself. It does seem as though a vast expansion of literary criticism and literary journalism occurred at the very moment when literature was attempting to withdraw from the public arena.

The day after the seminar on history and literature, I gave a lecture at Pomona College on Franz Overbeck, Professor of Theology and Church History at Basle during Nietzsche's years as Professor of Classical

Philology there and Nietzsche's close friend. In his *Christlichkeit der heutigen Theologie,* published the same year (1873) as Nietzsche's second *Unzeitgemässe* on David Strauss—the two friends had the essays bound together and always referred to them as *"die Zwillinge"* ("the twins")— Overbeck developed the argument that the original Christian faith had nothing to do with theology. It existed out of time, as it were, and in indifference to the world and to history, since the early Christians believed the end of the world was at hand. Theology developed out of the need to adapt this timeless and ahistorical phenomenon to a historical world—the world that survived despite the expectation of its imminent end. Theologians are thus, in Overbeck's view, in the very act of trying to "save" Christianity and secure its survival in the world, its worst betrayers; they are "the Figaros of religion," "panderers coupling Christianity and the world," "old washerwomen drowning religion in the endlessly flowing stream of their chatter"—i.e. people who "translate" what is otherworldly and "accommodate" it to the world, make it *speak.* Many lovers of literature would claim that literature, as we moderns understand it, resembles religion in that it too does not "speak to us" and engage with us in the manner of worldly discourses like politics or ethics or history—or literary criticism; that one does not argue with a work of literature; and that the proper way of engaging with it is not to subject it to scholarly and historical analysis and criticism, but to let ourselves be "inspired" by it—as the believer is supposed to be inspired by and to imitate the example of Christ. The opposition between *Bildung* and *Wissenschaft,* about which I have been concerned in my current research on anti-modernist trends in the culture of nineteenth century Basle, now strikes me as in many respects immediately pertinent to the difference I am suggesting between literature, on the one hand, and literary criticism or history, on the other. *Bildung,* from Humboldt to Burckhardt, was seen as an internalization and appropriation of creative capacities, the reward of a labor of imagination. *Wissenschaft,* in contrast, resulted from analysis and criticism; it was a product of reason.

In any account of the relation of literature and literary criticism that aims to preserve what might be called the "sacred" character of literature, its independence of both history and reasonable communication, I and those like me—critics, teachers, professionals, interpreters and communicators—can expect to be associated with democracy and "improvement," with "modernism," as Overbeck and Nietzsche understood that term. Most probably it will be found that there is something middle-class or "middle-brow" about us: that we are noisy brokers of culture, merchants of words and ideas, officious managers of a vast "demi-culture," which its detractors denounce as the deadly enemy of genuine culture. On the other hand, the writers and artists, the true believers, will be viewed or will view themselves as supremely otherworldly, indifferent to, even contemptuous of all adaptations and interpretations of what they do, and of all the activity of the lilliputians trying to tie them down, exploit them, make them manageable, marketable, exchangeable, integrate them into the "system," as we say, and thereby relieve them of whatever might be effectively disturbing or overwhelming in their work.

The situation I am evoking was outlined in the late twenties of the present century by Albert Thibaudet, one of the most brilliant French literary critics and scholars of his time:

> The Republic of writers will not be the Republic of professors. The Republic of writers is on the Right, today at least, along with the Republic of economics. The Republic of professors is on the Left, along with the Republic of politics. Like the Republic of economics, the Republic of writers, or rather the Republic of Letters, places a premium on *production*. At its limits lie the apothesosis of genius, awe and reverence in the face of its unlimited rights, the imperialism of intellect. Like the Radical Party politician, the Republic of professors places a premium on *distribution*. A good average remains the ideal of the class.[6]

Of course, it is possible to argue, as Marivaux did, against the view of literature as a kind of sacred text, different from ordinary discourse. The Norwegian scholar Stein Olsen, for example, presents a view of literature as a "social institution of the same kind as an economic system, defined by a normative structure which makes possible a literary practice." Literature, Olsen points out, is "obviously a social practice in the minimal sense that it involves a group of people among whom literary works are produced and read." But it is also one, he claims, in a stricter sense:

> i.e. as a practice whose *existence depends* both on a background of concepts and conventions which create the possibility of identifying literary works and provide a framework for appreciation, and on people actually applying these concepts and conventions in their approach to literary works. If literature is such an institution then aesthetic judgement must be understood as defined by the practice and apart from the practice literary judgements are impossible. And a literary work must then be seen as being offered to an audience by an author with the intention that it should be understood with reference to a shared background of concepts and conventions which must be employed to determine its aesthetic features. And a reader must be conceived of as a person who approaches the work with a set of expectations defined in terms of these concepts and conventions. Somebody who did not share this *institutional background* would not be able to identify aesthetic features in it because he did not know the concepts and conventions which define these features.[7]

---

[6]*La République des Professeurs* (Paris: Grasset, 1927), 235. The Republic of letters that "today, at least," as Thibaudet prudently phrased it, is on the Right, has thus shifted ground considerably from that of the seventeenth and eighteenth centuries, of Bayle, and Fontenelle, and Voltaire. That Republic of Letters represented precisely a Utopian ideal of community and exchange among the educated, the *"philosophes."*
[7]Olsen, *The End of Literary Theory,* 12.

When I came across it quite recently, Olsen's account of literature struck me as strikingly similar to the account I give of history in the essay. "The institutional approach to literature rests," he writes, "on an assumption of a fundamental agreement concerning what literature is and what literary judgments are. The task of literary aesthetics is to display the nature of this agreement."[8] Olsen's position appears to lend support to the idea of a community of literary practitioners and scholars not unlike other professional communities. It also seems to me compatible with the idea of literature as a special sphere of activity cut off from and even opposed to ordinary communication, to which I alluded earlier, even though it also suggests that literature's withdrawal has been only relative. Literature has not, after all, been "saved" from the world: it has simply been evacuated, like religion, to a special sphere. What Olsen's account leaves out, however, is an element that some people still believe to be essential to literature: its power, on privileged occasions, to re-invade the world or to draw its readers out of it and affect them deeply. As is well known, many of the revolutions in esthetic "paradigms" have occurred precisely in order to restore this almost magical power of the literary text, in order to liberate it from the "bureaucratic" responses of trained and practised readers such as Olsen supposes we all have to be. While I am highly sympathetic to Olsen's basic argument, in short, it does seem to me that his account of literature is subject to the same criticism as my account of history. It is so eager to establish community and responsibility and norms of conduct that it does not leave room for what disturbs the norms and in the end renews them.

I have not altered my commitment to a rational historiography. I am simply more ready than I was when I wrote the essay to acknowledge that, while it is vital for a civilized community to establish and follow norms of conduct, it cannot altogether rule out the unexpected, the unorthodox, the improbable and implausible. In the end, however, it remains the community that determines whether a new practice or a new paradigm will be accepted as part of what Olsen calls the "concepts and conventions" that underlie all social institutions.

One of the central questions of both the essay on historiography and my investigation of the "anti-modernists" of Basle, it now appears to me, is the status of scholarship (and that means also, in large measure, my own status). What does it mean to be a critic, an "interpreter," a "professor"? What does the professor profess? We are all in this together, I think—historians, literary critics and theologians. If I want to make my activity as rational as possible, to remove it from "art" and "magic" and "charisma"—as I do—am I thereby engaged in a Philistine effort to turn all culture into dead culture, so that it can be controlled and regulated? Is criticism, as poets have long maintained, the secret enemy

---

[8]Ibid., 13.

of literature, in the way that theology, according to Overbeck, is the "Satan of religion" or history, according to Nietzsche, the arch-foe of culture?

I admit that I am not sure how to answer. I am reminded, however, of Roman Jakobson's account of the way innovations occur in folklore works. An individual performer of a folklore work may give a novel and individual rendering of his or her model, Jakobson explained. But only those variations that the community finds acceptable will be integrated into the work and taken up by subsequent poet-performers.[9] The community of professional historians has similar powers, it seems to me, except that its members are expected to know the reasons for their decisions. In any social activity, in short, there is always a certain tension between the rules and conventions that make it possible to sustain the activity over time, as a shared activity, and the innovations which make it possible for it to develop and expand but which also threaten its survival and coherency. In some historical situations, however, the tension will be greater than in others; in some there will be a greater fascination with innovation as both desirable and menacing, in others a premium will be placed on conservation and norms. But no social practice seems thinkable which does not allow for both conservation and innovation.[10]

In addition to the members of the Claremont Colleges seminar, I am indebted for advice, criticism, or simply the opportunity to exchange ideas on questions of history, to friends and colleagues in the history department at Princeton, notably Natalie Z. Davis, Dan Rodgers, Jerry Seigel, and Laurence Stone. To the last two, who took the trouble to read the essay and comment on it in some detail, as well as to Robert Palmer, Fritz Stern, and Richard Vann, who were equally generous with their time and counsel, I wish to express special thanks. Finally, I would like to acknowledge the debt that all of us who are interested in historiography owe to Hayden White. The criticism of my essay that has most worried me is that, while there may be some point in making the argument I make in the context of current literary criticism, it does not make sense in the context of historical studies, where the danger is not— it is objected—an overemphasis on the linguistic and literary dimension of historical writing, leading to a crisis of faith in the nature of historical knowledge, but a continued, obstinate indifference to the work of White, Mink, and others who have tried to make historians more aware of what they are doing when they write history. I hope it is clear that

---

[9] Roman Jakobson and Piotr Bogatyriev, "Die Folklore als eine besondere Form des Schaffens," *Donum Natalicium Schrijnen* (Nijmegen and Utrecht, 1929), 900–913.

[10] On this point, see also Jan Mukařovsky, "The Esthetics of Language," in *A Prague School Reader on Esthetics, Literary Structure and Style*, ed. Paul Garvin (Washington, D.C.: Georgetown Univ. Press, 1964), 31–69, and my essay "Literature and Democracy," *MLN*, 1971, 86:761–89, reprinted in *Velocities of Change: Critical Essays from MLN*, ed. Richard Macksey (Baltimore: Johns Hopkins Univ. Press, 1974), 3–31.

without White's work, my own makes no sense. My object has not been to excuse or justify self-satisfaction, and my best interlocutor is the historian who has followed White and the whole inquiry into history-writing that his work has stimulated. White was in fact one of the first to read my essay, in an earlier version that I was required to submit to the organisers of a conference on History and Literature we both attended at Dartmouth College in the spring of 1985 (under the auspices of the School of Criticism and Theory). I doubt that he accepted my argument, but he recognized the spirit in which it was made, and judged it with his usual sympathy and generosity. My essay is directed not against him, as some may have thought, but to him.

I. A PERSONAL NOTE

Man's respect for knowledge is one of his most peculiar characteristics. [. . .] But what distinguishes knowledge from superstition, ideology or pseudoscience? [. . .] The demarcation between science and pseudoscience is not merely a problem of armchair philosophy: it is of vital social and political relevance.
—Imre Lakatos, *The Methodology of Scientific Research Programmes* (Cambridge, 1978).

Y a-t-il, comme il le pensait, une vérité française et une vérité allemande? [. . .] Ou bien n'y a-t-il, dans l'ordre politique et moral, qu'une vérité? Barrès n'a jamais employé le terme de pragmatisme. Il n'en reste pas moins un des fondateurs et l'un des plus puissants vulgarisateurs de ce pragmatisme européen, dont les mystiques totalitaires tirent aujourd'hui les plus implacables conséquences.
—Albert Thibaudet, *Histoire de la littérature française de 1789 à nos jours* (Paris, 1936).

Plusieurs croient que le poète et l'historien soient d'un mesme mestier; mais ils se trompent beaucoup, car ce sont divers artisans qui n'ont rien de commun, l'un avecques l'autre.
—Pierre Ronsard, *La Franciade* (1572), Preface.

The view of historiography that I try to develop in the following pages is not one that I have espoused in the past. Implicitly at least, I have been close to the position with which I now take issue. I do not write, therefore, without sympathy for the ideas I now find problematical. As it would seem desirable that my own background and point of view should be as clear as possible, I propose to preface my essay with a brief narrative of my own interest, as a teacher of literature, in historiography.

The focus of my first inquiry into historiography (*Medievalism and the Ideologies of the Enlightenment*, published in 1968, but based on a D.Phil. thesis written under the late Jean Seznec a decade earlier) was not so much the historiographical text as the institutions and methods of historical scholarship and literary history in the seventeenth and eighteenth centuries. My aim was to "contextualize" the activity of scholarship and to establish its relation to contemporary ideologies and to the social and institutional conditions in which it was carried out. In the last chapter I touched on the different ways in which history was understood and used in the Enlightenment and in the Romantic period, and it was to nineteenth century historiography that I turned shortly afterwards, focusing now more on narrative historical texts than on historical scholarship and attempting something like a literary analysis of those texts. I had already, in 1963, published a short article on Voltaire's *History of Charles XII* in which I tried to demonstrate that, though individual constituents of the narrative had been borrowed from oral and written testimonies and from other historical narratives, after being subjected to the careful scrutiny expected of a modern writer and a *philosophe*, the narrative itself was essentially literary and rhetorical. The *History of Charles XII* was not, I maintained, substantially different in design from Voltaire's better known *contes*, and, as with the *contes*, its meaning was an effect of the narrative design rather than a rational deduction from the "facts" of the case. The articulation of this argument was rather crude. But my basic aim was already, I think, to challenge, indirectly at least, the institutionalized boundary between "history" and "literature" (which in any case dates only from the last century) by showing that historical discourse is subject to the same kind of analysis as any other discourse, and that historical narrative has much more in common with "fictional" narrative than historians are normally willing to allow, in short that history—at least in so far as it takes the form of a text, especially a narrative text—is not a science in the naive sense in which that term might still have been understood by some historians two decades ago. To a large extent, this is what I have tried to do in subsequent studies of Michelet (1974 and 1985), Augustin Thierry (1976), and, with less success I fear, Gibbon (1982).

The complement to the project of reading historical texts as literature has been a continued commitment to a historical approach to "literature." For this reason, I have always wanted to reconsider as "history" the very historical texts that I had analyzed as "literature." There has never been any question, for me, of subordinating "history" to "literature." But just as historical narratives are not transparent representations of historical reality, the historical meaning and testimony of literary texts is not to be found, as I see it at least, in their passively "reflecting" reality, but in their structuring of it, in their rhetoric or the relation they establish with their readers, and in the different ways literature itself has been institutionalized. Besides attempting "literary" analyses of historical texts, therefore, I have also tried to argue that the categories of "literature" and "history" have a history (which in turn can be represented only through the mediation of literary and rhetorical patterns), that literature as a social institution and, above all, as a subject of instruction in schools and an instrument of cultural formation and communication, is part of history and subject to historical analysis, and, more specifically, that the idea of "literature" underwent a significant transformation in the early nineteenth century.[1]

---

[1]"History and Literature: Reproduction or Signification," in R. H. Canary and H.

Some time in 1972 the manuscript of Hayden White's *Metahistory* was placed before me as a member of the editorial board of the Johns Hopkins University Press. On the chapters devoted to Ranke, Marx and several others I felt I was not fit to pronounce, but I was dazzled by the important programmatic opening section of White's book on the "Poetics of History." It seemed to provide a theoretical foundation and perspective—of which I myself would have been incapable—for the kind of work I was engaged in. I still remember a decisive lunch with the editors and my colleague Maurice Mandelbaum, whom I revered not only because of his sharp intelligence and benign irony, and because he was a real philosopher (whereas I only dabbled in philosophy and had signally failed to convince my teachers in the moral philosophy class at Glasgow University that I had any aptitude for the subject whatsoever), but because of the infinite kindness he had shown me since the day I arrived in Baltimore in the fall of 1958 and the friendship that had grown up between us. Maurie said he didn't agree with White, but that he thought I probably did and that I should make the decision. Not much more was to be prized out of him. With something close to terror I said I thought we should publish it. It was probably the best decision I made during my term on the board of the Hopkins Press. After *Metahistory* I discovered the marvellously lucid essays of the late Louis Mink on the role of narrative in historical writing, and it has been, on the whole, from a narrativist point of view that I have taught my own classes in historiography.

Though my assumptions have been basically narrativist, however, I have never myself ventured any general theory of historical knowledge or historical discourse. This is only partly attributable to the philosophical inadequacy I have already acknowledged. I think I also had misgivings about some of the suppositions and implications of the narrativist position. In the early seventies I was keenly aware that the mixture of Lukacs and Sartre that had served as the basis of my understanding of the relation between history and literature throughout most of the 1960s would have to be reviewed in the light of the challenge from structuralism and a revived formalism. But because of that background in Lukacs and Sartre, I was never altogether comfortable with the extreme idealism, relativism, and formalism that often accompany narrativism.

The problem was and remains a practical one for me as a teacher of literature, in as much as I have been unwilling to isolate "literature" from "history." Juxtaposing different types of text—say Descartes, Corneille, Richelieu, and the Preface to the *Dictionnaire de l'Académie française*—and trying to show that they may be carrying and promoting

---

Kozicki, eds., *The Writing of History: Literary Form and Historical Understanding* (Madison: Univ. of Wisconsin Press, 1978), pp. 3–39; "Literature and Education," *New Literary History*, 1982, 13:341–71.

the same ideology, one is at least dealing with comparable items. But how does one establish a relation between a rhetoric or an ideology and historical reality (not an account of class conflict, for instance, but actual class conflict)? When literary scholars set a text in historical context, as some of us are prone to do, what are we actually doing? What is the context? How do we know it? Is it in some way ontologically prior to the text under review, and indeed to any text, as was once unreflectingly assumed? Is there, in short, "another side" of stories; or in reaching out beyond stories, do we simply abut on other more comprehensive stories, as seems to happen all the time in Diderot's *Jacques le fataliste?* And if so, what are our own explanations and interpretations but new stories constructed out of bits and pieces of already existing stories? The issue—which no literary scholar interested in the political and historical dimension of literature can escape—was clearly put by Hayden White:

Within a long and distinguished critical tradition that has sought to determine what is "real" and what is "imagined" in the novel, history has served as a kind of archetype of the realistic pole of representation [. . . .] Nor is it unusual for literary theorists, when they are speaking about the "context" of a literary work, to suppose that this context, the "historical milieu," has a concreteness and an accessibility that the work itself can never have, as if it were easier to perceive the reality of a past world put together from a thousand historical documents than it is to probe the depths of a single literary work that is present to the critic studying it. But the presumed concreteness and accessibility of historical milieux, these contexts of the texts that literary scholars study, are themselves products of the fictive capabilities of the historians who have studied those contexts.[2]

White's position is not popular among historians, as one can easily imagine. Nevertheless, though most "practising" historians profess to distrust him or try to ignore him, a number have, quite independently it seems, expressed views that are at least compatible with his. In a striking development of the strict professionalism he defended in the "Personal Retrospect and Postscript" of his *Reappraisals in History* (London: Longmans, 1961), Jack Hexter, for instance, has adopted a surprisingly ironical view of his activity as a historian. In his own historical writing, Hexter declares, he is interested in presenting a model of experience, not just an account of the facts. The story he tells, he says, is indeed limited by the rules of the historical game, a code of behavior that all historians subscribe to, but his ultimate goal is to tell a story similar to that of the fictional writer. Of one of his own works, Hexter writes that if he has failed to convey in it what he calls a sense of "triumph and tragedy," then he has failed altogether.[3] Pushed to the

---

[2]"The Historical Text as Literary Artifact" in *The Writing of History,* ed. R. H. Canary and H. Kozicki (Madison: Univ. of Wisconsin Press, 1978), 42–43. White has maintained this position with absolute consistency. See his "The Question of Narrative in Contemporary Historical Theory," *History and Theory,* 1984, 23:1–33, at pp. 19–21.

[3]*The History Primer* (New York, 1971), 207–8.

limit, Hexter's position might turn out to be perilously close to that of Maurice Barrès, who once remarked impatiently to a friend: "My book on Persia is already done . . . The only trouble is that I have to go to the damn place—to satisfy a bunch of idiots" [i.e. philistine bourgeois readers who expect a travel narrative to "reflect" a concrete experience of the place in question].[4] Hexter's comments and—more provocatively—those of Barrès highlight the problem of the cognitive content of historical texts, the relation between the meaning of a historical narrative, something it shares with fictional narratives, and the empirical facts it refers to. What is it, they provoke us into asking, that orders the historical text and governs its meaning? Is it the "reality" to which the individual statements in the text ostensibly refer? Or is it a principle of literary composition? How, in short, do the idea and order of history which we derive from historical texts, and to which, if we are literary scholars, we refer the texts (including the historical texts) that we study, relate to historical "reality"? Is "the past as history" no more, in Valéry's pithy phrase, than "a piece of imagination based on records"?[5]

In addition, the debate about literature and history seems deeply embedded in an institutional rivalry of which no literary scholar can remain long unaware. When the modern study of literature was institutionalized in the last century it was quite clearly associated with and subordinated to history. The "other" of literature was then, thanks to the Romantics, not history but rhetoric. The present organization of the discipline in the university (by national languages and historical periods) reflects both that association and that subservience. What seems to have been happening lately is a kind of *Umwertung* of these values (though it might be more accurate to speak of a Restoration). The languages of philosophy and history, which for so long have sat in judgment over all other forms of discourse, are themselves being put in question. The judges are being judged. Visibly, it is the renewed contemporary emphasis on language and rhetoric, not as ornaments of texts, but as their deep structure—and I would add, as institutions in which social meanings and cultural patterns are already inscribed—that has produced the discomfort presently felt both by historians who worry about the epistemological foundations of their discipline and by literary critics who have been accustomed to thinking of history as a kind of *garde-fou* of the imagination.

No doubt this development can be seen as a logical consequence, in the long run, of the shift from the Enlightenment vision of history as a system of explanation, on the model of Newtonian physics, to the

---

[4]Reported in Jérôme et Jean Tharaud, *Mes Années chez Barrès* (Paris: Plon, 1928), 204–205.

[5]"Unpredictability" [1944], in Paul Valéry, *History and Politics,* trans. Denise Folliot and Jackson Mathews (New York: Bollingen Foundation, 1962), 69. [*Collected Works of Paul Valéry,* vol. 10]. See likewise "Historical Fact" [1932]: "The past is an entirely mental thing. It is nothing but images and beliefs. Notice that we use a kind of contradictory procedure for evoking the various figures of the different epochs. On the other hand, we need the free use of our ability to pretend, to live other lives than our own; on the other, we must restrain that freedom in order to take account of documents" (ibid., 122–23).

Romantic vision of history as hermeneutic, on the model of Biblical studies and the new philology. Reality itself came to be viewed, from this Romantic perspective, as a "text" to be interpreted.[6] More immediately, however, the re-examination of historical discourse, and with it of the validity of studying literature in historical context, appears to have been provoked by two influences in particular. The first is that of the Anglo-American school of analytical philosophy of history, represented by the work of Danto, Gallie, Mink, and the two Whites (Morton, and later Hayden). The implication of the analytical philosophers' view of narrative form as constitutive of meaning in history is that the meaning of the historical work, as of any text, is discovered not in its referentiality but in its textuality. The second is that of French and American "deconstructionism," with its emphasis on the virtually uncontrollable signifying power of language, which overwhelms all attempts at "scientific" discourse. The deconstructionist perspective privileges literary discourse over all others as the only one that is not deluded, the only one that knows its true character is to be untrue. Not surprisingly, self-consciousness and irony are seen as the distinction and the defining characteristics of literature.

The current trend in the philosophy of history toward some version of narrativism—in Britain and the U.S. it has been going on for some time; in France it goes back probably to Barthes' famous essay of 1967; in Germany it seems to have happened with great suddenness in the last decade or so—is probably also part of a general revision of the view of scientific knowledge and rationality that we have held in the West since the seventeenth century philosophers undermined the validity of arguments based on authority and consensus and imposed a new ideal of certainty based on the twin criteria of systematicity and external correspondence with the structure of reality. The ideal of the mirror of nature, as is well known, was unfavorable to rhetoric. As the art of persuasion, rhetoric dealt, by definition, with what was not clearly and distinctly true. Moreover, as the repository of *opinio,* natural language— in contrast to a lost "original" language or an artificially constructed "philosophical" or "scientific" language—was perceived to be an obstacle to that direct perception of incontrovertible truths on which any valid knowledge was supposed to rest.

The combined influence of Anglo-American analytical philosophy of history and Franco-American "deconstructionism" has ensured that what was an avant-garde opinion when it was proposed by Hayden White—basically, that history is a linguistic and rhetorical artifact constrained by a genre rule specifying reference to conventionally agreed upon historical "facts," that "fiction," in other words, informs

---

[6]See Jürgen Kempski, "Aspekte der Wahrheit," especially section 2: "Die Welt als Text," in his *Brechungen: Kritische Versuche zur Philosophie der Gegenwart* (Rohwolt, 1964), 278–94; also Gossman, "History as Decipherment: Romantic Historiography and the Discovery of the Other," *New Literary History,* 1986, 18:23–57.

"history"—has become the orthodoxy of today, the common coin of popular journalism and undergraduate essays.[7] In a recent number of the *Princeton Tory*, for example, a bright undergraduate undertook the task of "Saving Faith Philosophically" on the strength (though hardly in the spirit) of Richard Rorty's highly successful critique of epistemological foundationalism and the model of the mirror of nature. "True" and "false," this young author announced with a blandness and aplomb that would surely have dumbfounded Nietzsche or Wittgenstein, are "dispensable metaphors." Similarly, a rhetorical view of history is simply assumed by the author of a *New York Times* review of the revised edition of Leon Edel's famous biography of Henry James:

Biography, as everyone knows, is a species of imaginative writing that tempts even the most exact chronicler to aspire toward the visionary freedom of the novelist. We no sooner discern a shape in someone's history than we impose our own design on the flux of life [. . . .] We decide, with a necessary arbitrariness, that [. . .] this life, like a novel, had a "plot" with dips and peaks, and even a "theme." We select from our materials what conforms to these and discard what does not. More, we let ourselves speculate about things invisible and unrecorded [. . .] in which we are willing to believe so our story will make sense. If we are imaginative writers, we make our story persuasive; perhaps, after all, it is true"

—and so on.[8] I recall that when I read this review—and there are hundreds like it—I felt a mixture of sympathy, irritation, and guilt. Sympathy for what still seems to me to be the kernel of truth in it, irritation at the trivializing and banalizing of what is, after all, a disquieting problem, and guilt that I have at times participated in this kind of talk and added to its volume.

In addition, I have become increasingly apprehensive that, as in the 1920s and '30s, the platitudinizing of formerly avant-garde positions is undermining the critical and rational tradition that still provides the best defense of democratic societies and is encouraging the kind of irrationalism, the cult of violence, that in the past has benefited the revolutionary Right far more than the Left. It is noteworthy that German scholars and philosophers (I am thinking particularly of Hans Albert and Ernst Topitsch) are especially sensitive to this aspect of "postmodernism." Indeed, my own misgivings have been expressed more effectively than I could have expressed them myself by a young German

---

[7]See Johan Goudsblom, *Nihilism and Culture* (Oxford, 1980). Goudsblom emphasizes the pervasiveness of what he calls "nihilism" in our culture. A hundred years ago, he claims, the "nothing is true" theory was the prerogative of an elite. Today it is so pervasive that "the problematic can no longer be called a personal acquisition; it is a platitudinous clincher, casually adopted and indiscriminately used. In many of the forms in which nihilism is avowedly manifest, one would have to go a long way to find the truth imperative [which, for G., is what underlies nihilism] for quite different influences are here at work" (p. 190).

[8]Michael Berger, "Saving Faith, Philosophically," *The Princeton Tory*, 1985, 2,1:30–36; *New York Times*, 24 November 1985.

historian, one who writes frequently on historiography and is by no means closed to modern reflection on the subject. According to Jörn Rüsen, the present re-examination of historiography

coincides with a notable turn in the value system of cultural discourse. The Enlightenment tradition, with its dominant concepts of critical reason and emancipation, is losing its power to carry conviction and, in striking analogy to the culture criticism of the late nineteenth century, irrationalism and mythical thought are acquiring more and more literary, academic, and journalistic prestige. Nietzsche is the leading figure of a post-structuralist intellectual avant-garde that is fascinated by the destruction of reason [. . . .] The weaker the conviction among historians that their intellectual activity is or at least should be, rationally informed, the more easily historiography is made over into an instrument of ideology. The much celebrated revival of narrative in histori-ography erodes that conviction, as does the much discussed metahistorical thesis of the essentially rhetorical character of historiography. The new irratio-nalism finds allies in those historians to whom the task of interpreting the historical experience of what men have done to each other and to their world in the course of time in terms of rational criteria has come to seem too difficult. The rationality of historical thinking fades when historians give up on providing historical memory, through their work of interpretation, with a glimmering of discernible and intelligible coherency. In such conditions Clio, their muse, might well reach for the greasepaint of irrational meanings as a cover for meaningless-ness. In this way, history gets decked out as a poetic activity. (It is not an accident that Theodor Lessing's well known *Geschichte als Sinngebung des Sinnlosen* has been recently reissued) [. . . .] History as a scholarly activity rests on rational principles that have been achieved and elaborated as a result of a long historical process. That process is called "emancipation." In our time, that notion of emancipation, together with the Enlightenment tradition on which it is supported, has a bad press. It is up-to-date and certain to win the applause of the post-modern intellectual avant-garde to unmask whatever is connected with emancipation and Enlightenment as illusion, and to celebrate the counterhistory to the history of modern emancipation as the one by which the contemporary situation is determined [. . .] History must resist the temptation to swim with the tide of the new longing to forget the horrors of the present in the arms of the past. Instead, it should constantly remind us of those historical experiences that are intimately connected with the currents of irrationalism, the effort to bring the civilizing culture of Modernity to an end, and the creation of myths that promise to release us from the burden of discursive thought.[9]

---

[9]"Historische Erinnerung und Menschliche Identität—praktische Wirkungen der Historiographie," *Universitas* 1984, 39:393–400, at pp. 393–94, 394–95, 400. On the contemporary criticism of reason, see also A. Gargani, ed., *La Crisi della ragione* (Turin: Einaudi, 1980), and F. Ferraresi, "Julius Evola: Tradition Reaction and the Radical Right," *Archives Européennes de Sociologie*, 1987, 28:107–51.

II. THE NECESSITY OF NARRATIVE

Suspicion of narrative attended the beginnings of modern historiography in the late seventeenth century and it has never altogether abated.[10] The great age of historical narrative, usually celebrated as the heyday of "history" in general, may in fact be something of an aberration. It is at least arguable that Bayle's influential *Historical and Critical Dictionary* is more exemplary than any of the Romantic narratives. And in this founding work, not only is the choice of the dictionary form and the conventional alphabetical arrangement of the material a striking repudiation of narrative, the challenge to narrative is repeated in each individual article. The thin narrative line of each entry (often a bare line or two at the top of the folio page) is invariably interrupted by innumerable note signs referring the reader to the copious scholarly discussions that fill up the space below, question the reliability of the narrative, and generate further distracting notes in the margins. It is virtually impossible to read Bayle for the narrative without feeling the ground give way under one's feet.[11] As he himself liked to say, "On ne sait où planter les pieds." In his preface, moreover, Bayle makes his distrust of narrative explicit by citing a perfectly coherent but factually "false" narrative sentence ("Coriolanus's mother obtained of him what he refused the sacred College of Cardinals and the Pope himself, who went to meet him"), in order to demonstrate that such a sentence may be as meaningful and as rhetorically effective as any

---

[10]Michel de Certeau, the author of *L'Ecriture de l'histoire,* argues that the critical position of the late seventeenth century scholarly historians is the essential determination of modern history. The distinction of "fiction" and "history," in other words, is the condition of the existence of what we understand by history (*La Philosophie de l'histoire et la pratique historienne aujourd'hui,* [Ottawa, 1982], 19).

[11]Even the story of Uriel Da Costa (article "Acosta"), which could be read as one of the earliest narratives of the *philosophe* as hero and martyr, is put in question, since it is taken from Da Costa's own account of his life, and that, Bayle observes in his notes, was inevitably not disinterested; in addition, some of the most elementary facts about it, such as the date of Da Costa's suicide, are unresolved matters of scholarly dispute. If there is disagreement and uncertainty about an event that occurred within living memory, Bayle implies, if there is a space between the logic of narrative and the succession of events in such a case, what confidence can one have in accounts of events that are supposed to have occurred centuries, even millennia ago? Luc Weibel makes a similar point in his discussion of the article "Abelard": "L'insertion de la remarque dans le fil de la narration produit . . . deux effets; elle rompt la continuité, le 'nappé' du récit référentiel; elle fait vaciller son rapport avec la 'réalité' en renvoyant à un autre discours dont le sens est incertain, et en rompant la connivance de la logique du récit avec la succession temporelle des événements" (*Le Savoir et le corps: Essai sur le Dictionnaire de Pierre Bayle* [Lausanne: Editions de l'Age d'Homme, 1975], 59).

"true" narrative.[12]

Nevertheless, even Bayle, critical and anti-systematic as he was, could not conceive of history except as a narrative against whose seductions the honest historian must be eternally vigilant. If history was possible at all, it could only be as a narrative—one that historians constantly weave, unravel, and weave back again.[13]

Later Enlightenment critics of traditional historical narrative—and I am thinking much more of Malthus or Playfair or Eden than of Voltaire or Macaulay, who never really did much to avoid it—could also not escape narrativity. Though they are intended to establish correlations, Eden's elaborate tables of incomes and prices both imply and support a narrative of the *changing* state of the laboring people of England and of the *evolution* of English society from Norman times to those of the historian himself at the end of the eighteenth century. Likewise, nonverbal, predominantly graphic representations of historical data—an invention of Englishtenment scholars who wanted to avoid rhetoric and natural language and, in Playfair's words, speak "a language that all the world understands," thus ensuring that the information, from which they hoped future generations would be able to derive the *laws* of historical development, "should go down in such a form and manner as that any person might, even though a native of another country, understand the nature of the business delineated"[14]—even such graphic representations, familiar to us from historical atlases and the tables accompanying modern historical texts, can only be interpreted by being translated into the form of a narrative. Minard's *carte figurative* of the gradual disintegration of the Grande Armée in the course of the Russian campaign of 1812-1813, first published in 1861 and celebrated on account of its clever representation of multivariate data, in fact "tells a rich coherent story," as one scholar has said, "[. . .] far more enlightening than just a single number bounding along over time."[15]

---

[12]Quoted from the 2nd English ed. (London, 1734). As Weibel puts it: "Ce discours parfaitement cohérent, parfaitement 'coulé', contient des énoncés dont les référents sont incompatibles, mais ces référents, extérieurs au discours n'en entravent pas le fonctionnement" (*Le Savoir et le corps*, p. 31). In part, Weibel contends, it was Bayle's publisher Leers who asked him, for the sake of sales, to "tempérer un peu son mépris pour les histoires qu'il a lues, et de faire que son Dictionnaire ne soit pas complètement un anti-dictionnaire, qu'il contienne également des récits et des fables comme le public les aime" (p. 40).

[13]Bayle liked to compare reason and philosophy to "une véritable Pénélope, qui pendant la nuit défait la toile qu'elle avait faite le jour" (*Dictionnaire historique et critique*, art. "Bunel, Pierre," note E; in another passage, which I have been unable to retrace, reason is "une coureuse qui ne sait où s'arrêter, qui, comme une autre Pénélope, détruit elle-même son propre ouvrage: *diruit, aedificat, mutat quadrata rotundis*"). The same could be said of Bayle's idea of the work of history.

[14]William Playfair, *The Commercial and Political Atlas, Representing by Means of Stained Copper-Plate Charts, the Progress of the Commerce, Revenues, Expenditure, and Debts of England, during the Whole of the Eighteenth Century* (3rd ed. London: J. Wallis, 1801 [1st ed., 1786], Preface, v).

[15]Edward R. Tufte, *The Visual Display of Quantitative Information* (Cheshire, Conn.: Graphics Press, 1983), 40.

Figure 1

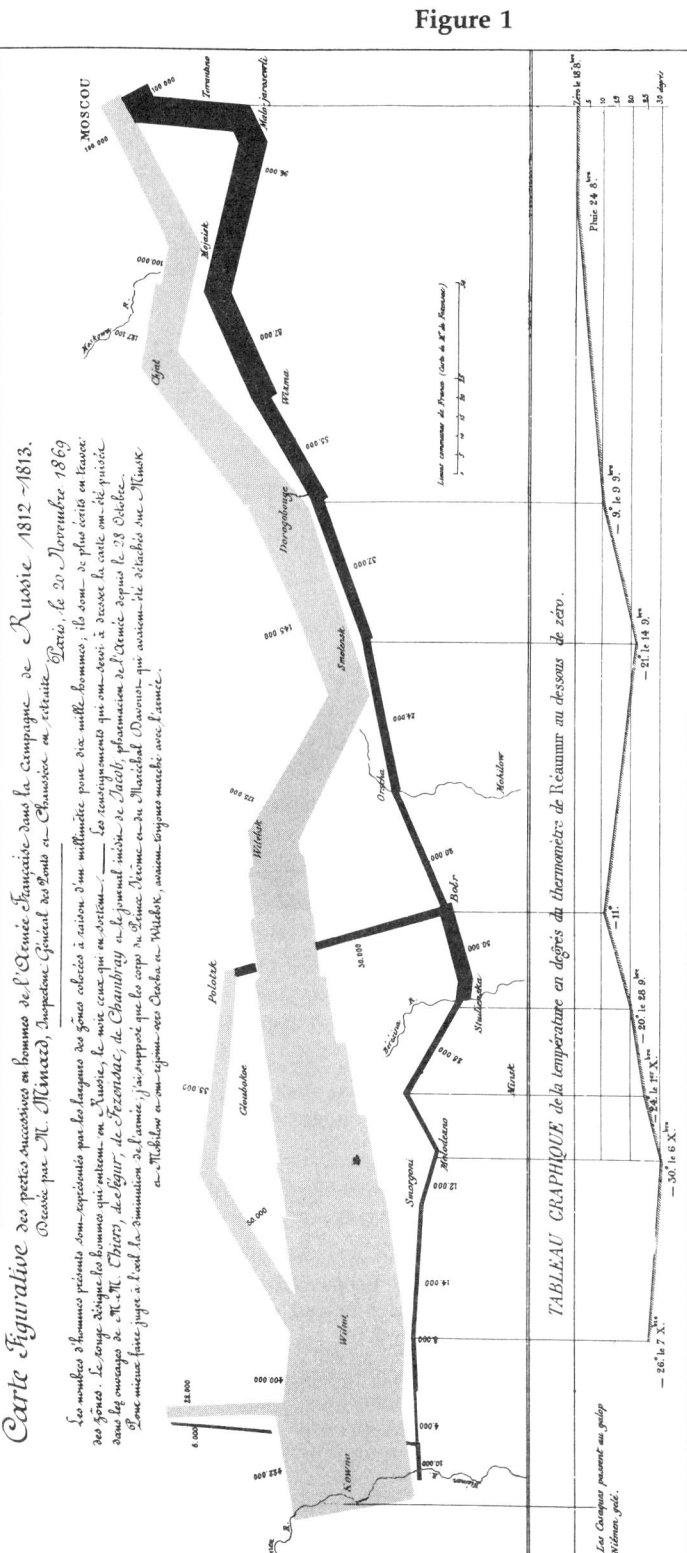

Napoleon's March to Moscow   The War of 1812

This classic of Charles Joseph Minard (1781–1870), the French engineer, shows the terrible fate of Napoleon's army in Russia. Described by E. J. Marey as seeming to defy the pen of the historian by its brutal eloquence, this combination of data map and time-series, drawn in 1861, portrays the devastating losses suffered in Napoleon's Russian campaign of 1812. Beginning at the left on the Polish-Russian border near the Niemen River, the thick band shows the size of the army (422,000 men) as it invaded Russia in June 1812. The width of the band indicates the size of the army at each place on the map. In September, the army reached Moscow, which was by then sacked and deserted, with 100,000 men. The path of Napoleon's retreat from Moscow is depicted by the darker, lower band, which is linked to a temperature scale and dates at the bottom of the chart. It was a bitterly cold winter, and many froze on the march out of Russia. As the graphic shows, the crossing of the Berezina River was a disaster, and the army finally struggled back into Poland with only 10,000 men remaining. Also shown are the movements of auxiliary troops, as they sought to protect the rear and the flank of the advancing army. Minard's graphic tells a rich, coherent story with its multivariate data, far more enlightening than just a single number bouncing along over time. Six variables are plotted: the size of the army, its location on a two-dimensional surface, direction of the army's movement, and temperature on various dates during the retreat from Moscow. It may well be the best statistical graphic ever drawn.

Edward R. Tufte, *The Visual Display of Quantitative Information*   Graphics Press Box 430 Cheshire, Connecticut 06410

More recently, the repeated attacks on narrative by members of the *Annales* school of historians, starting with the classic little book of Marc Bloch, have also not really been targeted at narrative as such. They turn out, on inspection, to be directed rather against a certain kind of narrative, one that has in fact been abandoned by many writers of fiction and even by some writers of history: I mean what we usually refer to as "classical" narrative, with its well defined characters, plot-line, and point of view—the "healthy" kind of narrative that Ortega y Gasset once contrasted with Proust's "sickly" descriptions of shifting psychological states, or that Voltaire had in mind when he criticized Montesquieu's *Considérations sur la grandeur et la décadence des Romains* for its provocative disdain of it.[16] It is narrative in this sense that the pioneers of the *Annales* school wanted to relegate to the dustbin of historiography, not narrative as such, about which most of them probably did not think very hard or at all.[17]

Some narrative framework seems to be implied in the very act of recognizing and identifying an individual historical fact. (Would a fact which was identified in terms of a non-narrative system of relations be recognizable as a *historical* fact?) It is only by being recognized as part of a potential narrative, in other words, that historical material becomes meaningfully historical. For example, Collingwood's hypothesis (based on the deciphering of tombstone inscriptions) that parties of Scots had settled down peacefully in Southern England at the time of the Roman occupation—which Leon Goldstein cites as an example of scholarly

---

[16]Letter to Thieriot, November 1734, Besterman #D-803.

[17]Since history never really escaped from narrative, talk of a return to narrative, now quite frequent among historians (see Laurence Stone's "The Revival of Narrative: Reflections on a New Old History," *Past and Present*, 1979, 85:3–24) probably refers to a return to traditional narrative forms. That may be good or bad. On the face of it, I would suspect that it was not a promising move, if indeed it exists. As Siegfried Kracauer once pointed out, the popularity of biography, with its strong and clear narrative line, was a negative reaction, in the interwar years, to the difficulties of modern life and of modern views of reality presented in the complex, self-conscious, and fragmented narratives of the best fictional writing of the time. It would be a pity if history were to serve such a function today. Fortunately, that does not appear likely. At least one of the examples given by Stone, Duby's *Le dimanche de Bouvines: 27 juillet 1214* (Paris: Gallimard, 1973), is anything but a straightforward old-fashioned narrative. On the contrary, Duby unpacks the simple narrative of Rigord, the old chronicler, subjects it to a thorough critical analysis, and prises out of it a richly revealing account of medieval society and values. Bouvines, in Duby's own words, "est un lieu d'observation éminemment favorable pour qui essaie d'ébaucher une sociologie de la guerre au seuil du XIII siècle dans le Nord-Ouest de l'Europe." The traces it has left "instruisent sur le milieu culturel au sein duquel l'événement vient d'éclater, puis survit à son émergence" (p. 13). Even the way Bouvines is interpreted as an episode in subsequent narratives of the history of France, from the time immediately following the event down to the last century, is used by Duby as a source of historical information about the values and shared representations of a historical community. This is not a return to "histoire événementielle" by any stretch of the imagination. On the whole, I think Hayden White would agree with my judgment of the *Annales* school's criticism of narrative history; see his "The Question of Narrative in Contemporary Historical Theory," *History and Theory*, 1984, 23:1–33, at pp. 8–10.

reconstituting of the historical record in contrast to narrative art[18] – is itself not a single nugget of fact, but a small narrative, which in addition implies a larger context or argument in which it takes its place and in light of which the "raw" data were themselves perceived. This point is made vividly by Martin Rudwick in his fascinating study of a major debate in nineteenth century geology. Rudwick remarks that the empirical observations of the geologists were by no means direct or uninformed by theory. "When examining an exposure of rock on the seashore or in a road cutting, a geologist did not have the same perceptions as a nearby fisherman beaching his boat or a villager passing in a cart. What the geologist perceived were the already interpreted features of strata with a measurable orientation, containing identifiable fossils, and capable of being integrated on the spot – however provisionally – into an imagined picture of vastly larger structures and sequences."[19]

The narrative structure of historical reasoning survives the most brutal assaults on its actual verbal texture, as a glance at the four or five line entries in publications such as *Historical Abstracts* will demonstrate. The fact that historical accounts weather such treatment more successfully than literary narratives suggests among other things that the time of narration and of reading may be of less importance in a historical narrative than in a literary one. The performance of the narrative, in other words, the reader's actual experience of it, may be less significant in the case of historical accounts than the achieved restructuring of the reader's idea of (some part of) the world. Given the intellectual or cognitive, rather than mimetic or esthetic ambition of most historiography, this seems hardly surprising. Perhaps it is what Louis Mink had in mind when he wrote that stories

aim at producing and strengthening the act of understanding in which actions and events, although represented as occurring in the order of time, can be surveyed as it were in a single glance as bound together in an order of significance, a representation of the *totum simul* which we can never more than partially achieve [. . . .] In the understanding of a narrative the thought of temporal succession as such vanishes – or perhaps, one might say, remains like the smile of the Cheshire act.[20]

It seems to me difficult to avoid the conclusion that narrative is an essential and not an accidental characteristic of historiography, despite the persistent – and probably necessary – suspicion of it among historians and their constantly renewed attempts to escape its constraints

[18]Leon J. Goldstein, *Historical Knowing* (Austin and London: Univ. of Texas Press, 1976), 201.
[19]*The Great Devonian Controversy: The Shaping of Scientific Knowledge among Gentlemanly Specialists* (Chicago and London: Univ. of Chicago Press, 1985), 431.
[20]"History and Fiction as Modes of Comprehension," in *New Directions in Literary History*, ed. Ralph Cohen (London: Routledge and Kegan Paul, 1974), 120.

and routines. One must, I think, concur with Jörn Rüsen's cautious statement of his own position in the matter: "Historical accounts are not securely based for the simple reason that the events they describe really happened. The rigor of an historical account does not depend exclusively on its empirical content, it also derives from the narrative organization of that content."[21] Rüsen and his colleague Hans Baumgartner have also argued that description and explanation in historiography turn out on examination to be not equivalents of or alternatives to narrative, but subordinate, in any historical account, to a basic narrative scheme, so that "narration is a structural concept of the historical object, of historical objectivity, and not merely one form of representation along with others."[22] This means that historical works which at first sight might appear to deliberately eschew narrative, such as Tocqueville's *The Ancien Regime and the Revolution,* Burckhardt's *Civilization of the Renaissance,* Fustel's *Ancient City,* or in more recent times, Laslett's *The World we have Lost* (or, to mention also a fine but not especially celebrated work of recent academic historiography, Yves-Marie Bercé's *Révoltes et Révolutions dans l'Europe moderne, XVI-XVIII siècles* [Paris, 1980], which is a comparative study of early modern revolts), do in fact have a narrative structure, however deeply embedded. Similarly, the most elementary "facts" of history—the fall of Constantinople, the discovery of America, the demographic upswing in eighteenth century France, the industrialization of Japan under the Meiji—are always apprehended as narratives, composed of other, "smaller" facts, arranged in a particular yet variable way. And these facts in turn prove to be themselves more or less complex narratives, be it a single "event" or act of the French Revolution, or the demographic pattern of a single parish.

---

[21]"Erklärung und Theorie in der Geschichtswissenschaft," *Storia della Storiografia,* 1983, 4:3–29.

[22]*Erzählforschung: ein Symposium,* ed. Eberhard Lämmert (Stuttgart: J. B. Metzler, 1982), 697 ("Erträge der Diskussion" of Day 4 of the symposium, devoted to "Erzählung und Geschichte").

III. THE PROBLEM OF INCOMMENSURABILITY

What disturbs many practicing historians, as well as some who, like myself, have argued for greater recognition of the literary and rhetorical aspects of historiography, is less the claim that historical explanation must assume narrative form than the argument for the incommensurability of historical narratives that often accompanies that claim. That historical narratives are incommensurable with each other appears to follow from a combination of two prior arguments. The first is that no narrative is a simple reflection or copy of past reality, since past reality is by definition no longer present. As Chaim Perelman wrote in one of his invaluable essays on historical and legal argument,

What the historian is given is never the past, which, by definition, is no longer. What he is given, the objects he must respect and may not deform, are the present sources of historical knowledge. In fact, the historian's task is, through his narrative, to reconstitute the human past in an intelligible and impartial manner, on the basis of as scientific a study of the sources capable of throwing light on that past as possible [. . . .] No historian can go down into the past like a deep sea diver to describe what he sees. We can know the past only from the traces of it that remain.[23]

It is impossible, therefore, to evaluate different narratives of the "same" occurrence or situation, since the narratives cannot be compared with the occurrence or the situation they are supposed to represent. In fact, to the degree that the narratives actually construct the occurrence, it is probably not strictly correct to speak of the "same" occurrence. Between the signifier and the referent, there is the signified, and the relation of the signified to the referent, especially to an absent referent, must be highly problematical. Peter Munz has described this position very effectively: "We can compare one clock to another clock, but we cannot compare any clock to time and it makes therefore no sense to ask which of the many clocks we have is *correct*. The same is true of any story, including historical narratives. We cannot glimpse at history. We can only compare one book with another book."[24] Munz still writes here of

[23]"Objectivité et intelligibilité dans la connaissance historique" (1963), in his *Le Champ de l'argumentation* (Brussels: Presses universitaires de Bruxelles, 1970), 361–71.

[24]P. Munz, *The Shapes of Time* (Middletown, Ct.: Wesleyan Univ. Press, 1978), 221. See also pp. 16–17: "Whereas we can translate a photograph into a painting and a painting, taking its life into our hands, into a verbal statement and an English text into a Russian text, we cannot translate what actually happened into anything. We can translate what

25

comparing different narratives, as did W. B. Gallie, one of the early narrativists:

The kind of explanation that I claim to be characteristic of histories cannot be confirmed, or even preferred against other possible explanations, except via the acceptability of the narrative which it enables the historian to reconstruct or resume. If the narrative has now been made consistent, plausible, and in accordance with all the evidence, if it is the best narrative that we can get, then the explanation that helped us to get to it is the best explanation as yet available.[25]

The criteria to be used in the comparison of different narratives are thus internal consistency or "plausibility" and conformity "with all the evidence."[26] But as many people, including myself, have pointed out, evidence only counts as evidence and is only recognized as such in relation to a potential narrative, so that the narrative can be said to determine the evidence as much as the evidence determines the narrative. The relation between empirical evidence and theory is surely no clearer in history than it has turned out to be in the natural sciences — a point to which I shall have occasion to return later. As one critic of Gallie remarked, "That the narrative must be 'in accordance with all the evidence' is something anyone who chooses to write about history is expected to say, but it is not easy to determine from the passage — or from the book from which it is taken — how the nature of historical evidence functions in Gallie's thinking about history."[27]

Nevertheless, however hard it may be to determine on what grounds one narrative is to be judged preferable to another, the possibility — in principle — of making comparisons and of their having some bearing on the validity of the historical work *as knowledge* is denied neither by Gallie nor by many other narrativists. The view that historical narratives cannot be measured against historical reality does not, apparently, make them incommensurable with each other. On the contrary, for Gallie and others, they can be evaluated only by being compared with one another.

It is when the anti-realism of the narrativists is complemented by a

---

somebody *thought* happened into another language and seek to establish equivalences between different media — at least up to a point. But we cannot translate reality; for to do that we would have to have a picture of or a text about it in the first place [. . .] But the ineluctable truth is that there is no face behind the mask and that the belief that there is, is an unsupportable allegation. For any record we could have of the face would be, precisely, another mask." Similarly Dande Cohen, "Structuralism and the Writing of Intellectual History," *History and Theory*, 1978, 17:175–206: "Historians 'touch' the object of their discourse only by recourse to the already-meant; hence every shred of historical meaning belongs to the discourse and not to the objects [. . . .] There is no direct communication possible between the referents of historical discourse and the discource about a referent."

[25] *Philosophy and the Human Understanding* (London: Chatto and Windus, 1964), p. 124.

[26] With modifications Gallie's view seems to have won broad acceptance. See, for instance, a recent study by the Dutch scholar F. R. Ankersmit: *Narrative Logic: A Semantic Analysis of the Historian's Language* (The Hague, Boston, London: Martinus Nijhoff, 1983).

[27] Goldstein, 99.

second claim—namely that the shape of historical narratives is determined by rhetorical tropes, which are themselves not subject to historical investigation but function rather in the manner of Kantian categories—that historical narratives come to seem thoroughly incommensurable. Though few eighteenth century historians or critics ever lost sight of the relation between rhetoric and history (Gibbon and Blair can both be usefully consulted on this point), and even certain nineteenth-century criticis, like Sainte-Beuve, noted that historical narratives are much affected by literary and rhetorical patterns,[28] the systematic elaboration of a modern poetics of history has been principally the achievement of Hayden White. White's view is essentially that

---

[28]The experience of 1848 appears to have made Sainte-Beuve sensitive to the element of chance in history and to have encouraged him to criticize the Romantic providentialist view, in which everything seems in retrospect to have been inevitable. Guizot's philosophy of history, in particular, is "trop logique pour être vraie. Je n'y puis voir qu'une méthode artificielle et commode pour régler les comptes du passé" (review of Guizot's *Discours sur l'Histoire de la Révolution d'Angleterre,* added to the 1850 edition, in *Causeries du lundi* [1857], vol. 1, pp. 311–31). Whatever doesn't fit the historian's pre-arranged plot is discarded, Sainte-Beuve complains: "Toutes les causes perdues [. . .] sont considérées impossibles, nées caduques, et de tout temps vouées à la défaite. Et souvent à combien peu il a tenu qu'elles ne triomphassent." In the end, the historian acts as a kind of Providence, imposing order on the chaos of historical reality: "L'histoire [. . .] vue à distance, subit une singulière métamorphose, et produit une illusion, la pire de toutes, celle qu'on la croie raisonnable. Dans cet arrangement [. . .] qu'on lui prête, les déviations, les folies, les ambitions personnelles, les mille accidents bizarres qui la composent et dont ceux qui ont observé leur propre temps savent qu'elle est faite, tout cela disparaît [. . . .] On ne juge plus que de haut. On se met insensiblement en lieu et place de la Providence. On trouve à tout accident particulier des enchaînements inévitables, des nécessités, comme on dit." The historian, ultimately, is an artist: "L'homme, il faut bien se le dire, n'atteint en rien la réalité, le fond même des choses, pas plus en histoire que dans le reste; il n'arrive à concevoir et à reproduire que moyennant des méthodes et des points de vue qu'il se donne. L'histoire est donc un art; il y met du sien, de son esprit, il y imprime son cachet, et c'est même à ce prix qu'elle est possible" (essay on Mignet, in *Portraits contemporains,* vol. 5, pp. 225–56, at p. 240). Or again: "L'histoire n'est pas un miroir complet ni un fac-simile des faits; c'est un art. L'histoire, quand on parvient à la construire, est un pont de bateaux qu'on substitue et qu'on superpose à cet océan dans lequel [. . .] on se noierait sans arriver" (review of Saint-Priest, *Histoire de la Royauté, Portraits contemporains,* vol. 4, pp. 1–30, at p. 3). So it happens that two histories of the "same" event may well appear to the reader as descriptions of quite different events. Reading Guizot's account of the English Revolution, Sainte-Beuve relates, "je me suis donné le plaisir de lire en même temps des pages correspondantes de Hume: on ne croirait pas qu'il s'agisse de la même histoire!" (review of Guizot, p. 321). Sainte-Beuve's preference for Hume is a signal of the retreat, after 1848, from Romantic historiography to what White would call the "ironical" mode characteristic of Burckhardt: "Ce que je remarque, surtout, c'est qu'il m'est possible, en lisant Hume, de le contrôler, de le contredire quelquefois: il m'en procure les moyens, par les détails mêmes qu'il donne [. . . .] En lisant M. Guizot, c'est presque impossible, tant le tissu est serré et tant le tout s'enchaîne. Il vous tient, et vous mène jusqu'au bout, combinant avec force le fait, la réflexion et le but." It is worth noting that, as the social conflicts of his time came to appear more and more intractable and as the hollowness of the July Revolution, from which so much had been expected, became increasingly obvious, even Michelet, the arch-Romantic, began to fear that the imposing and orderly edifices of science and historical knowledge might be only fragile, man-made constructions built over a terrifying chaos of often cruel realities, the order of which, if there is one, is neither intelligible to man nor related to his welfare. See my "History as Decipherment" (note 6 above) and "Jules Michelet and Romantic Historiography," in *Scribner's European Writers* (New York, 1985), vol. 5, pp. 571–606, at pp. 598–99, 602.

history is "ultimately determined by formal and rhetorical structures, most fundamentally the tropes of metaphor, metonymy, synecdoche, and irony, and that it can escape the impasse it now allegedly finds itself in only by ironic consciousness of its own formal nature, that is by accepting its similarity to fiction."[29]

Distinguishing between different aspects of the narrative text of history ("chronicle," "story," and "plot" in White's terminology, the latter two being roughly equivalent to "fable" and "subject" in the terminology of the Russian formalists), White argues that "the events reported in a novel [i.e. the elements of the "fable"—L. G.] can be invented in a way that they cannot be (or are not supposed to be) in a history." Because of this, it is easier to distinguish "chronicle" and "story" or "fable" in a work of history than in a literary fiction. For

unlike the novelist, the historian confronts a veritable chaos of events already constituted, out of which he must choose the elements of the story he would tell. He makes his story by including some events and excluding others, by stressing some and subordinating others. This process of exclusion, stress, and subordination is carried out in the interest of constituting a story of a particular kind.[30]

Now if it is true that the historian simply *uses* materials haphazardly thrown to gether in a repository conventionally designated as history, in order to construct his narratives according to the same rules as the writer of fiction, then indeed different histories are incommensurable. To a narrativist like Gallie, historical narratives do propose an analysis of past reality, even if they cannot be held up and measured against that reality. From White's perspective, the historian is not even offering an analysis of past reality (even though the historian himself may not know this). The kind of understanding he provides is no different from that provided by the writer of fiction. Histories are thus incommensurable in the way that novels are. Two different accounts of what we conventionally designate as "the French Revolution" (but what is "the French Revolution," the rhetorician would ask, except what it is figured as in the narratives that purport to "describe" or "analyze" it?)[31] are in fact simply two different stories, whose materials happen to have been selected from those usually placed in a bin marked "French Revolution." In a short essay written for George Iggers's and Harold Parker's

---

[29]The summary of White's position is by Suzanne Gearhart, *The Open Boundary of History and Fiction* (Princeton: Princeton Univ. Press, 1985), 7.

[30]*Metahistory: The Historical Imagination in Nineteenth Century Europe* (Baltimore: The Johns Hopkins Univ. Press, 1973), 6, note 5. The proposition that the historian confronts a "chaos" of events—not very different in the end from Sainte-Beuve's view (note 28 above)—seems to me quite questionable in fact. The historian does not normally confront such a chaos of events. Events are almost always encountered in more or less intelligible patterns and relations (narratives), which the historian may wish to challenge, reject, or alter. He does not work alone, without predecessors.

[31]White: "Historians *constitute* their subjects as possible objects of representation by the very language they use to *describe* them" ("The Historical Text as Literary Artifact" in Canary and Kozicki, pp. 41–62, at p. 57).

International Handbook of Historical Studies Louis Mink formulated the incommensurability argument with his customary force and clarity:

If the cognitive content of written histories is in part exhibited in its form (over and above that part of its cognitive content which consists of the referential meaning of its factual statements), then it is difficult to see in what sense two different histories can be said to agree or to be incompatible with each other, since complex forms cannot be restated as propositional assertions, which alone can be comparable or incomparable. The response of what might be called the New Rhetorical Relativism in historiography to these considerations is that indeed narrative syntheses are cognitively incommensurable with each other.[32]

In a later essay, Mink reaffirmed this assessment of the consequences of White's "poetics" of history. Though it contains relativism within the limits set by the four tyes of emplotment, Mink wrote, *Metahistory* "specifically rejects any possibility of a rational choice among them. Of course, something accounts for the historian's choice of one mode of emplotment rather than another, but what that choice expresses is both extra-historical and extra-philosophical; it represents an esthetic or political preference, as a matter of individual taste or commitment."[33]

---

[32]"Philosophy and Theory of History," in *International Handbook of Historical Studies* (Westport, Ct.: The Greenwood Press, 1979), 17–27, at p. 25.

[33]"Is Speculative Philosophy of History Possible?" in *Substance and Form in History: Festschrift for W. H. Walsh*, ed. L. Pompa and W. H. Dray (Edinburgh: Edinburgh Univ. Press, 1981), 107–19. White himself makes his position perfectly clear: The works of the great historians "cannot be 'refuted,' or their generalizations 'disconfirmed,' either by appeal to new data that might be turned up in subsequent research or by the elaboration of a new theory for interpreting the sets of events that comprise their objects of representation and analysis" (*Metahistory*, p. 4); "This is not to say that we cannot distinguish between good and bad historiography, since we can always fall back [sic!] on such criteria as responsibility to the rules of evidence, the relative fullness of narrative detail, logical consistency, and the like to determine this issue. But it is to say that the effort to distinguish between good and bad interpretations of a historical event [. . .] is not as easy as it might at first appear, when it is a matter of dealing with alternative interpretations produced by historians of relatively equal learning and conceptual sophistication. After all, a great historical classic cannot be disconfirmed or nullified either by the discovery of some new datum [. . .] or by the generation of new methods of analysis" ("The Historical Text as Literary Artifact," 59). It is worth observing that the terms of valuation here ("good" and "bad", rather than "true" and "false" or "valid" and "invalid") themselves direct the reader toward ethical and esthetic, rather than intellectual judgment, to the existential rather than the rational.

IV. ROMANTICS AND POSITIVISTS

The incommensurability argument has created few ripples among those who actually *do* history. In general, historians tend to go about their business casting only an occasional, indifferent glance in the direction of philosophers and literary critics. In this respect they are not very different from natural scientists. Many of them probably hold intuitively to some form of realism. But even those among them who think and write *about* history have reacted less vigorously than one might have expected. As I suggested earlier, the pervasiveness of the "nothing is true" theory, to which the writings of the later Wittgenstein, Thomas Kuhn, and Michel Foucault, to say nothing of Nietzsche himself, have imparted an air of respectability and familiarity, seems to have dampened traditional responses. I know of one brave effort to resurrect historical realism, but to the best of my knowledge it has found few takers.[34]

A more plausible defense of historical knowledge against the narrativist erosion of it has been attempted on the basis of the old distinction between historical writing and historical research. Leon Goldstein's *Historical Knowing* (Univ. of Texas Press, 1976) is interesting because it upholds the "scientific" character of historical research without being in any way committed to realism. On the contrary, Goldstein unequivocally adopts a constructivist position:

As much as we may want to say that a true account of some past event is true in virtue of the fact that it accords with what actually took place when the past was present, we have no way to make that belief operative in historical research [. . . .] What we know about the historical past we know only through its constitution in historical research, never by acquaintance.[35]

The very choice of "knowing" rather than "knowledge" for the title of Goldstein's book is probably in itself significant: history, it is suggested, denotes a way of knowing as much as a content of knowledge.

The argument developed by Goldstein is similar to that put forward by Murry G. Murphy in *Our Knowledge of the Historical Past* (Indianapo-

---

[34]Adrian Kuzminski, "Defending Historical Realism," *History and Theory*, 1979, 17:316–49.

[35]Leon J. Goldstein, *Historical Knowing* (Austin and London: Univ. of Texas Press, 1976), xix–xxi. See also p. 136: "Nothing appears in the practice of history that corresponds to the object of witnesses. The past—real or historical—certainly does not. The witness confronts the object; his account is the result of his encounter with it from his own perspectives. The historian in no way confronts the *real* past. And rather than confront it, he constructs the *historical* past."

lis, 1973). The relation of historical fact to historical evidence is not one of inference: "Rather the historical occurrence is hypothesized in order to make sense of the evidence."[36] Goldstein is thus by no means a historical realist. Indeed, his conviction that inquiry and research rather than narration are at the heart of the historian's activity leads him to the view that only what has been reconstituted as a result of the historian's critical analysis and evaluation of evidence is properly historical knowledge. "Historical knowing is a way of knowing not by acquaintance."[37] In conformity with this definition of historical "knowing" as a way of knowing rather than a content of knowledge, "Goldstein specifically rejects the material of memory as properly "historical knowledge. Memories of lived experience, he maintains, including the memories and written narratives of historian-participants must be treated as documents and evidence, not authorities, and "they must be subject to the same sort of critical examination that a properly trained historian applies to all of his evidence."[38] Goldstein's own authority here is Collingwood and he refers explicitly to the latter's conception of history as "a certain kind of organized and inferential knowledge," quite distinct from memory which "is not organized or inferential at all," that is, as an autonomous discipline or practice which is characterized by the adoption of recognized methods and criteria.[39]

"Presence" is thus in no way an ideal for Goldstein. On the contrary: All knowledge, as he sees it, implies alienation. Goldstein's rejection of realism, which is no less radical than that of present-day narrativists, has nothing to do with narrativism, however. Indeed, if anything, it has been the partisans of narrative history, history *ad narrandum,* in opposition to the intellectualist history *ad probandum* of the Enlightenment, who, in the past at least, have wanted to recreate the living reality of history in the reader's imagination, to *resurrect* it, in Michelet's phrase. According to many historians, the function of the story in history and the special gift of the great history teacher are precisely that—to reawaken the dead, make the past speak, and put the reader or listener in its living presence.[40] Admittedly, such a claim is not compatible with modern narrativist theories such as those of Gallie or White. The idealism of the Romantics was not, as with White, a rhetoric, but a

---

[36]Goldstein, 127.

[37]Goidstein, 144.

[38]Goldstein, 147. See also pp. 156–57.

[39]Goldstein, 146.

[40]See, for instance, Goldwin Smith on the great history teacher in "The Gates of Excellence," in Goldwin Smith, ed., *The Professor and the Public: The Role of the Scholar in the Modern World* (Detroit: Wayne State Univ. Press, 1972), pp. 13–42: "When Professor Bossenbrook talked about the trade routes to the East you could almost see the sails of the great spice ships and hear the camels grunt by the wells of Trebizond" (p. 27). "Many of us have read accurate but boring books by authors who have never lifted their eyes from the documents, have never wondered if George Washington liked kidney pie . . ." (pp. 36–37). Not surprisingly, the historian is characterized by Smith as "the high priest of continuity" (p. 40).

metaphysics, a philosophy of universal analogy, according to which the mind can grasp the world, the self the other, because they are structured in the same way. To the Romantic historian, the imagination provided a true insight into the nature of historical reality, and it was this insight that made it possible to select, group, and interpret evidence. The modern narrativist espouses no such philosophy, and no claim to provide true knowledge, in the sense of "corresponding to historical reality," is made for history as he describes it.

Nevertheless, it is worth noting the explicitly anti-Enlightenment and conservative ideological stance of many of the former partisans of history *ad narrandum*, among whom one must count both Ranke and his wayward pupil Burckhardt.[41] To the nineteenth century historicists, the function of history was not to establish a critical distance between the present and the past, but to reconcile them and reestablish continuity between them. The famous phrase, not often quoted in full, in which Ranke disclaims any ambition to treat history in Enlightenment style as *magistra vitae*—"The task of judging the past and instructing the present for the benefit of the future has been ascribed to history: the present essay does not aspire to such high office: it aims only to show what actually happened" (*Geschichte der romanischen und germanischen Völker*, Foreword)—may be slightly disingenuous in the way it presents narrative continuity, and thereby the continuity of history, as somehow self-evident, simple, and unreflected, in contrast to the artificial and contrived nature of the "lessons" that the Enlightenment historians sought to draw from the facts.[42] Different as they are from their Romantic predecessors many modern versions of narrativism continue the Romantic opposition to the rationalist, "scientific," or nomothetic ideal of historical study that was common to Enlightenment scholars such as Malthus, Süssmilch, or John Millar, and extend it to the revised versions of that ideal that survive among Marxists as well as positivists.

At any rate, it is on the basis of a rejection of historical realism, which he shares with modern narrativists that Goldstein proceeds to build a case for history as an investigative activity, a method of establishing what he likes to call "the historical record," rather than an imaginative or poetic activity, a way of arranging "materials" in a meaningful order. The essential task of history for him is research, not writing: it concerns the establishing of facts, not meanings. Constructivist as he is,

---

[41]See Jacob Burckhardt, *Judgments on History and Historians*, trans. Harry Zohn (Boston: Beacon Press, 1958), 242 (Part V, The Age of Revolution; section 122, "German and French intellectual development in the eighteenth century").

[42]On the politically conservative character of the opposition to Enlightenment historiography of Ranke and the German historicist tradition in general, see Fritz Fischer, "Aufgaben und Methoden der Geschichtswissenschaft," in Jürgen Scheschkewitz, ed., *Geschichtsschreibung* (Düsseldorf: Droste Verlag, 1968), 7–28. See also on the deeply anti-Enlightenment tradition in German theories of knowledge, especially in the social sciences, Wolf Lepenies, *Die Drei Kulturen: Soziologie zwischen Literatur und Wissenschaft* (Munich and Vienna: Hanser, 1985), 245–66 et passim.

Goldstein would reject out of hand White's often repeated assertion that all history is philosophy of history.

Yet the two views of history, that defended by Goldstein and that defended by the narrativists, do not so much exclude each other as divide the territory of history between them. While the narrativists concern themselves with the way historical narratives shape the "raw materials" of history into configurations or stories and thus create and communicate meaning (in other words, with *Auffassung* and *Darstellung*, the third and fourth stages respectively in Ernst Bernheim's classic *Lehrbuch der historischen Methode* [Leipzig, 1889]), those who emphasize historical scholarship concentrate on the "logic of discovery" (*Heuristik* and *Kritik*) and leave the writing of historical narratives out of account as a matter of esthetics or ideology.

Thus, on the one hand, Hayden White concentrates on "historians and philosophers of distinctively classic achievement, those who still serve as recognized models of possible ways of conceiving history." The status of these writers as possible models of representation, he explains,

does not depend upon the nature of the "data" they used to support their generalizations or the theories they invoked to explain them; it depends rather on the consistency, coherence, and illuminative power of their respective visions of the historical field. This is why they cannot be "refuted," or their generalizations "disconfirmed," either by appeal to new data that might be turned up in subsequent research or by the elaboration of a new theory for interpreting the sets of events that comprise their objects of representation and analysis. Their status as models of historical narration and conceptualization depends, ultimately, on the preconceptual and specifically poetic nature of their perspectives on history and its processes."[43]

On the other side, Goldstein dismisses the narrative part of history disparagingly as mere superstructure, icing on the cake—"that part of the historical enterprise which is visible to nonhistorian consumers of what historians produce."[44] Strikingly, he seems not in disagreement with those who claim that historical narrative is subject to unchanging rhetorical categories, by which it is *prefigured:* "The superstructure has had comparatively little history [. . . .] We find a greater similarity between the way historical accounts were constructed in antiquity and the way such things were done in later periods than between, say, the science of the Greeks, that of the Renaissance, and that of our own times."[45] To the degree that history is a science that has developed its techniques and expanded its field of investigation, then, it is one at the level of what Goldstein calls its infrastructure—"that range of intellectual activities whereby the historical past is constituted in historical research." It is at this infrastructural level that there has been an expansion

---

[43]*Metahistory* (Baltimore: The Johns Hopkins Univ. Press, 1973), 4.
[44]Goldstein, 141.
[45]Goldstein, 141.

and refinement comparable to that which has occurred in the natural sciences. "The very [. . .] domain of historical evidence has been expanded from the reports of eyewitnesses, to which the ancient historians were largely limited, to the wide variety of things from which present-day historians have learned to extract such a variety of historical truth."[46]

Certain features of Goldstein's point of view would no doubt have appealed to the pioneers of modern historiography. To Marc Bloch, it will be remembered, history had "grown old in embryo as mere narrative" and had only just emerged as a "newcomer in the field of rational knowledge." Bloch placed his hopes for history as "a science in its infancy" in the expansion both of the range of evidence available to the historian and of the techniques for "cross-questioning" it. For Bloch appears to have believed, like Goldstein, that the application of new evidence and of new techniques of analysis, themselves borrowed from burgeoning new human sciences such as economics and demography, would ensure the "progress" of historical science. "Our history need be no more like that of Hecataeus of Miletus than the physics of Lord Kelvin or Langevin is like that of Aristotle."[47] Until recently at least, the same optimism motivated almost all the great figures of the *Annales* school. Thus for Pierre Chaunu, "no need to insist: quantification, the very condition of the coupling of history with the social sciences, is the condition *sine qua non* of any expansion of knowledge."[48] Ultimately, Chaunu argues, it will be possible to extend quantitative methods to the realm of behaviors and "mentalités" and to extract a society's attitudes to life and death from computerized studies of language and vocabulary, so that our knowledge of these things will finally cease to be subjective and impressionistic.

Two distinct views, then, of what constitutes the work of history. But a surprising measure of agreement between them. Rejecting narrative as inessential to the historical enterprise, Goldstein contends that narrative autonomy, the cognitive incommensurability of different historical narratives, is inseparable from the narrativist position. He is critical of Gallie's argument, referred to earlier, that no historical explanation can be preferred to another except "via the acceptability of the narrative that it enables the historian to reconstruct," that is to say, except by considering to what extent the competing narratives are "consistent, plausible, and in accordance with the evidence." Goldstein comments severely:

---

[46]Goldstein, 141–42.

[47]Marc Bloch, *The Historian's Craft*, trans. Peter Putnam (Manchester: Manchester Univ. Press, 1954), 13, 66–69, 21.

[48]"Conjoncture, structures, systèmes de civilisations," in *Conjoncture économique, structures sociales: Hommage à Ernest Labrousse* (Paris and The Hague: Mouton, 1974), 21–35, at p. 30.

That the narrative must be 'in accordance with all the evidence' is something anyone who chooses to write about history is expected to say, but it is not easy to determine from the passage—or from the book from which it is taken—how the nature of historical evidence functions in Gallie's thinking about history. But this apart, the words that tell [. . .] in our present context are those which proclaim that no explanation in a work of history may 'be preferred against any other possible explanations.' The reason for so striking a position is that, in Gallie's view, what determines or motivates any explanation is the role it plays in the narrative account in which it appears. Each narrative, having been produced in isolation from any other purporting to deal with the same theme, contains nothing that may be understood as in conflict with—or in any other way related to—narratives produced by other writers. Each work of history is to be judged only with reference to its own coherence and in isolation from everything else.[49]

According to Goldstein, the incommensurability thesis is exactly what makes at least the more radical forms of the narrativist thesis, however consistent, unacceptable as an account of what historians do and what history is about. In his view, historians do agree, disagree, and debate, and the working out of their differences is an essential moment in their constitution of the "historical record." The kind of agreement and disagreement he is interested in, he makes clear, is not that which "has to do with such matters as the significance of some course of events or the ways in which some historical occurrence is to be interpreted," it is not the kind of disagreement that "takes place during that phase of [the historian's] work which follows the stage at which the historical past is being constituted." Such disagreements, he concedes, may well tell us "about the character of the intellectual climate within which debate among historians takes place, about the ideological conflicts of the time, and about unresolved issues of theoretical social science. But they do not tell us anything about history qua that practiced discipline of which the purpose is to constitute the historical past."[50]

The disagreements that interest Goldstein are those that occur not at the level of large-scale interpretations of the meaning of events, but at the lower level at which historians try to determine "what different parts of the past must be like given the evidence."[51] At this level—that of the "facts" of history, or of what he himself refers to as "the historical record"—he insists both on the fact of disagreement and on the possibility of resolving it. If I may interpret this a little, I would say that the fact of disagreement underscores his contention that history is the product of research and discussion of evidence, not of a direct and incontrovertible vision of reality, that "the historian in no way confronts

---

[49]Goldstein, 99, quoting Gallie, *Philosophy and the Human Understanding* (London: Chatto and Windus, 1964), 124.
[50]Goldstein, 95.
[51]Goldstein, 137.

the real past" as a witness confronts an object, and that "rather than confront [the *real* past], he constitutes the *historical* past." The disagreement of historians, in short, is not like the disagreement of witnesses; it cannot be resolved by confronting their testimonies with the real past. On the contrary, "if the real past played any sort of role at all, then the sort of disagreement we are considering could not be possible."[52]

On the other hand, the possibility of reaching agreement, even if only provisionally, confirms Goldstein's basic contention that much historical disagreement is of a scholarly or scientific, non-ideological kind: it concerns the evaluation and working up of evidence, according to "methods and techniques generally agreed to." That is "what enables us to expect that, in the course of time, disagreements may be overcome and some tradition of scholarship established."[53] When we consider history as a professional discipline, the specter of skepticism is conjured. Disagreement among historians is essential. For Goldstein, disagreement in historical scholarship is similar to disagreement in the physical sciences. It has a dynamic rather than a static character, guaranteeing both the reality and meaningfulness of research, dialog, and exchange, and the impossibility of ever finding a definitive answer that can be checked against "reality."

To a narrativist like Hayden White, in contrast, disagreement among historians is less the sign of a lively ongoing process of communication and criticism in a common endeavor, as in the physical sciences, than an absolute condition. "The physical sciences," he writes, "appear to progress by virtue of the agreements reached from time to time among members of the established communities of scientists, regarding what will count as a scientific problem, the form that a scientific explanation must take, and the kinds of data that will be permitted to count as evidence in a properly scientific account of reality. Among historians, no such agreement exists, or ever has existed."[54].

One of Goldstein's illustrations of agreement having been reached and the historical record established tells a lot about the kind of problem

---

[52]Goldstein, 132.

[53]Goldstein, 132. In the case discussed by Goldstein, the point at issue, which it is hoped scholarly analysis and debate will settle, is the time and place at which the Dead Sea Scrolls were produced. Without some agreement among historians about "matters of fact" the discipline as a whole would be impossible, as the narrativists themselves would be the first to acknowledge. Importantly, however, changes occur concerning what is agreed upon, as sources and the techniques of investigating them are expanded and refined. Thus in his otherwise severely critical study of Fogel and Engerman, *Time on the Cross*, Herbert Gutman concedes that the old, ultimately racist view of slavery—that the slaves remained obstinately African, that they could not be transformed into diligent workers, and that the planters went broke trying to achieve this impossible transformation, so that slavery was fundamentally unprofitable—has been completely undermined by modern research. "It has been known for nearly half a century that slavery was profitable, and F+E deserve credit for reopening that question and focussing on the reason for its profitability" (*Slavery and the Numbers Game: A Critique of Time on the Cross* [Urbana, Chicago, and London: Univ. of Illinois Press, 1975], 169).

[54]*Metahistory*, 13.

he has in mind and the kind of solution he envisages. It concerns the deciphering of cuneiform by Sir Henry Rawlinson in the last century. Goldstein quotes from the memoir of Rawlinson written by his son:

From the ruins of a temple at Kileh Shergat, where researches were still being pursued, was exhumed a clay cylinder which "turned out to be a most valuable relic." It contained the annals of the first Tiglath-Pileser, a document of great length, belonging to a monarch anterior to the time of David in Israel, and by far the oldest historical inscription which had, up to that time, been discovered in the country. The cylinder reached Colonel Rawlinson in a very bad state, broken into fragments and in some parts pulverized. Colonel Rawlinson, however, succeeded in uniting the fragments with a composition of gum-water and powdered chalk, and obtained a copy of the entire inscription (with the exception of a few paragraphs), above 800 lines in length—a copy afterwards verified by duplicate cylinders, procured from the same mound, and in an almost perfect state of preservation. It was this inscription which afterwards played so important a part in the general verification of cuneiform interpretations, being simultaneously submitted for translation to the four chief experts, Sir Henry Rawlinson, Dr. E. Hincks, Dr. Jules Oppert, and Mr. Fox Talbot, who severally, without any communication, produced renderings which were substantially identical.

Goldstein makes two observations on this passage. Firstly, it allows us to appreciate Rawlinson's formidable command of Akkadian. For the task he confronted was far harder than the already difficult one of fitting together the pieces of a jigsaw puzzle, when one has no antecedent idea of the picture they are supposed to form after they have been reassembled. Since many of the fragments were not cleanly cut but worn, damaged, and incomplete, only his knowledge of the language enabled Rawlinson to bring off the reconstruction of the cylinder. Secondly, and more important, it was the coherence of the several translations that gave credibility both to the assumption that scratch-like marks on old baked clay were really human language and to the claim that some individuals had in fact learned to read that language.

What the Rawlinson case shows above all, for Goldstein, is both the inappropriateness of historical realism and the crucial importance of the commensurability of historical explanations. The realist would say that the cuneiform text is rendered accurately only if the scholar's present understanding of it accords with what the ancients who wrote it intended to express, just as the realist historian would claim that only an account of events that accords with the reality of the past is accurate. But such conformity cannot be realized either in the case of historical texts or in that of historical events. Goldstein maintains—and the point of the story and of the experiment it recounts is to have shown—

that we are confident that we have arrived at a notable degree of historical truth when those members of the historical community engaged in research on the subject in question reach a level of agreement [. . . .] The plausibility that accrues [to history] accrues to it, not because the results of its researches may be shown to satisfy the correspondence theory of truth [. . .] but rather [. . .] from the fact that many different scholars, by applying the techniques of the discipline to the body of so-called historical evidence, which in no way resembles events or historical facts, are able to achieve as broad an agreement as they actually have.[55]

The truth we can expect to achieve in history is thus similar to the kind of truth we get from an autopsy or an investigation of an air crash.[56]

In the last few years, my work on the maverick nineteenth century classical scholar, J. J. Bachofen, has made me increasingly aware of a disturbing kinship between many of the orthodoxies of modernism (perhaps postmodernism would be the better term) and the ideologies of the radical Right. In particular, as I observed earlier in this essay, I have been troubled by the ambiguous implications of the underlying estheticism of Nietzschean and many post-Nietzschean doctrines. The estheticizing of historiography is no longer something I view in the way I once did—as a liberation from a stuffy positivism. On the contrary, it now seems to me that it renders those who espouse it defenseless before the most dangerous myths and ideologies, incapable of justifying any stand. Gradually, I have begun to feel more sympathy with the point of view represented by scholars like Goldstein, who—as I read them at any rate—are trying to rehabilitate rationality and due process as essential moments of all our intellectual enterprises, without retreating to traditional metaphysical realism. Goldstein complains that "when philosophers discuss the problem of relativism in history [. . .] they are likely to talk about the ways in which the ideological orientations or commitment to values, from which there is no way to extricate the historian, lead to that state of affairs. But what we do not find are systematic discussions of the character of disagreement in history of the sort that would tend to illuminate the practice of history."[57] I find this remark well taken. As I indicated earlier, however, I am uncomfortable with Goldstein's rather sharp distinction between infrastructure and superstructure in history, and with his abandonment of the latter to ideology and esthetic preference. I do not believe the problems of historical scholarship can or should be so neatly distinguished from those of historical narrative. So I would like to venture further than he does and to plead that closer attention be paid to the structures of argument and debate at the level of the historical narrative itself. I would even propose a reconsideration of the possibility that the choice of narrative pattern may, perhaps should,

---

[55]Goldstein, 197–200.

[56]Goldstein himself sums up his position as follows: if we want to understand the character of historical thinking, "our attention must be riveted to the infrastructure of history and not to the superstructure. Before we can explain historical events, or weave them into the fabric of a narrative, we have to determine what they are [. . .] Historical thinking is that way of dealing with historical evidence so as to emerge with historical facts." (201)

[57]Goldstein, 95.

be determined not simply by ideology, esthetic preference, and the innermost preoccupations of the writer, but by a rational judgment about the capability of competing narratives and narrative strategies not only to account for existing evidence but to lead to the systematic, rather than merely haphazard or antiquarian uncovering of new evidence.

What gives me pause in Goldstein's position, in sum, is that it accepts and reinforces a division between judgment and decision, reason and *praxis*, that appears equally to underlie the position opposed to it. Goldstein's dichotomy of historical scholarship and historical narrative corresponds closely, in effect, to the opposition of positivism and existentialism that characterizes a great deal of modern thought. As Hans Albert has observed, however, that opposition rests in fact on a shared view of the world. The two orientations, he explains,

differ hardly at all on the dichotomy between knowledge and decision, but adopt radically different points of view in their evaluation of it. Whereas existentialism stresses decision, with its free and undetermined nature, emphasizes its irrationality, and declares scientific knowledge essentially uninteresting precisely because of its objectivity, positivism places the emphasis upon knowledge and objectivity, stressing its foundability and rational character, while dismissing decision and commitment to the realm of subjectivity and arbitrariness as philosophically uninteresting. One side seeks to eradicate objective knowledge because it allegedly fails to make contact with existence; the other seeks to avoid subjective decision because it appears to lie outside the sphere of rationality. However little they may have to say to one another, it is nevertheless clear that both movements start to some extent from common presuppositions. Both opt for a view in which rationality and existence part company, but one emphasizes the rational analysis of facts, while the other glorifies irrational existential decisions. Both are inclined toward a *facticist* conception of knowledge [. . .] Similarly both incline toward a *decisionist* treatment of value problems [. . . .] Any discussion between these two trends seems not merely unnecessary—since they have no need to dispute about the presuppositions they share—but actually impossible, because each party, by virtue of this shared ground, is obliged to concede to the other a sphere of influence that its own methods render inaccessible—the sphere either of 'pure' fact or of 'pure' decision.[58]

The modern opposition was already foreshadowed in a division between the empirical fact-gatherer and the theoretical thinker, the *érudit* and the *philosophe*, that seems to be as old as modern historiography—or modern science—itself, though of course the terms and the context of that earlier division were different from those in which the positivist was later pitted against the existentialist. I referred earlier to Bayle's effort to undermine traditional historical narratives by questioning their individual factual references. By the middle of the eighteenth century, the division between "facts" and "reasonings" led to

---

[58]*Treatise on Critical Reason*, trans. Mary Varney Rorty (Princeton: Princeton Univ. Press, 1985), 75–77.

skepticism about all historical knowledge that rested on traditional historical narratives, including those of the Ancients, and to the view, expressed on many occasions by Voltaire, d'Alembert, and Rousseau, that those traditional narratives are to be regarded as moral or political fables, whose value or moral "truth" is independent of their factual "truth." The truth of historical narratives which, in the medieval period had been on the whole internally a figurative and religious one and externally a matter of authority (a true story was one that was founded on an authoritative text), was now perceived as largely intelligible and philosophical.[59] In this respect it was no different from the truth of fictional narratives. "History is only a longer *conte*," as Voltaire's friend Cideville once declared.[60] Voltaire himself found it difficult to justify his relentless search for information about historical subjects. "I shall be like La Flèche," he used to say, referring to a character in Molière's *L'Avare*. "I shall turn everything to account."[61] In fact, though he continued to seek out more and more information, about Charles XII for instance, long after the publication of his history of that monarch, he was not led by any of it to make substantial changes to his original narrative, which in this instance is visibly the *amplificatio* of an initial antithesis or chiasmus, told in the comic mode, and based explicitly on a literary model: if the hero of the narrative imitates and seeks to rival Alexander, the historian himself seeks to rival Alexander's historian, Quintus Curtius.

Gibbon claims that Montesquieu had shown how the dross of erudition could be turned into history, and the study of facts reconciled with the activity of reason by the historian-philosophe.[62] The Romantics, however, rejected Montesquieu's mechanistic model of historical explanation (and along with it the metaphors he liked to draw from mechanics and hydraulics) in favor of an organicist one. Individual facts were revealed as rational and intelligible not by demonstrating their *lawfulness* in terms of a system of causes and effects (i.e., deriving the laws of history and society from facts and, in turn, testing these laws against the facts) but by discovering their *meaningfulness* as parts of a larger whole, at once figures of the whole of human history and moments in the process of its unveiling or revelation.

But the Romantic vision of history as a kind of sacred text to be

---

[59]On the changing range of meanings of the term "history," see the detailed study of Joachim Knape, *"Historie" im Mittelalter und früher Neuzeit: Begriffs= und gattungsgeschichtliche Untersuchungen im interdisziplinären Kontext* (Baden-Baden: Valentin Koerner, 1984); and the important entry by Reinhart Koselleck on "Geschichte, Historie" in Otto Bruner, Werner Conze, and Reinhart Koselleck, eds., *Geschichtliche Grundbegriffe: Historisches Lexikon der politisch−sozialen Sprache in Deutschland* (Stuttgart: Ernst Klett, 1975), vol. 2, 593-717.

[60]*Voltaire's Correspondence*, ed. T. Besterman, #6507 (1757).

[61]*Voltaire's Correspondence*, ed. T. Besterman, #860 (1735). The same expression recurs in a later letter of 1761 (Besterman #9285).

[62]*Essai sur l'étude de la littérature* (London: T. Beckett and P. A. de Hondt, 1761), 105.

interpreted by the methods of hermeneutic was no better able to effect a lasting reconciliation of research with narrative and of the individual fact with its meaning than the Enlightenment vision of history as a kind of machine to be understood by the methods of science had been. The danger for the Romantics, who always emphasized the quasi-divinatory powers needed by the historian in his work of decipherment, was that they would see too precipitously behind the fact to its meaning, that the meaning would overwhelm the fact. It was only to be expected that the rights of fact would be reasserted at some point, even at the expense of meaning, in a new positivism.[63]

I have been led to deal with an instance of the opposition between romanticism and positivism in my own work on the culture and society of nineteenth century Basle. According to J. J. Bachofen, the philologist, historian of law, and friend of Nietzsche during the latter's years at the University of Basle, there are two paths to knowledge: "the longer, slower, more toilsome path of rationally weighing and comparing evidence, and the shorter path which the imagination travels with the force and speed of electricity, when, excited by direct sight of and contact with the ancient sources, it grasps the truth in an instant, without having to pass through any intermediate stages."[64] Bachofen carried out his own work consistently with this view of scholarship. "He began," it has been said,

with a picture, an intuitive view, then collected evidence to support this [. . . .] The manuscript of *The Saga of Tanaquil* shows that the first draft of the book began with an imaginative sketch of Oriental society and then cited Tanaquil and other figures as illustrations of it, whereas the second (and published) draft begins with the story of Tanaquil and cites the other material as parallels in order to give the work 'a scholarly structure.' The published version was therefore a kind of fiction in that it did not reflect Bachofen's actual thinking. The material was selected to fit the theory and quite naturally some of it was not carefully examined.

The upshot is that "though Bachofen's theories were bold, based on brilliant use of evidence *(Kombinationen)* [. . .] he gave not the slightest attention to his sources, taking them all as equally valid. As a result some of his conclusions were remarkably accurate [by which one should no doubt understand "consonant with present scholarly opinion"−L. G.], some−such as relating the Sabines to Jews and Assyrians− downright absurd."[65]

Curiously enough, however, the views of Theodor Mommsen, whom Bachofen considered his arch-enemy and−as the leading exponent of the critical method and the scientific ideal in philosophy, which the

---

[63]See Sainte-Beuve's objections to early nineteenth century historians in note 28 above.

[64]J. J. Bachofen, "Selbstbiographie," in *Selbstbiographie und Antrittsrede über das Naturrecht*, ed. Alfred Baeumler (Halle/Saale: Max Niemeyer, 1927), 31.

[65]R. I. Frank, "Poesie: Mommsen versus Bachofen" (unpublished paper), pp. 9–11.

pious and conservative Bachofen identified not altogether erroneously with political radicalism—a truly Satanic figure, were in important respects not in contradiction with Bachofen's. After completing his *Roman History*, the brilliant literary success of which outraged Bachofen, Mommsen gave up large-scale narrative histories and from 1856 on devoted himself to exploring the relation between legal history and the history of the state, and to collecting, editing, and explicating texts and inscriptions. This made it possible for Arnold Toynbee, whose own historiographical practise was of course the opposite of Mommsen's, to make the outrageous quip that the great classicist's collected works are "like so many volumes of a learned periodical which happens to have had only one contributor."[66] The striking point is that Mommsen also believed that "phantasy [. . .] is the mother of all history as it is of all poesy" and that "the historian is much more an artist than a scholar."[67] Like Renan and other advocates of a positive, scientific method—even Weber to some extent—Mommsen appears to have felt that though science is clean and strong and masculine, there is something sterile about it. To go beyond what it can achieve, the more mysterious and more feminine powers of poetry and insight must be invoked. Mommsen's choice, like Weber's, was to renounce fantasy and narrative and confine himself to what could be argued and demonstrated "scientifically."

I am struck by the survival power of this division of historiography into the production of solid bricks by those one might be tempted to term the respectable bourgeois of the profession and the use of those bricks by "inspired" artists to create beautiful and seductive palaces of interpretation. In one of the most recent defences of narrativism the Dutch philosopher F. R. Ankersmit, referring in particular to Huizinga, recalls that in German and Dutch philosophy of history

a distinction is often made between "geschiedsvorsing" and "geschiedschrijving," i.e. between "historical research" and the "narrative writing of history." The term "historical research" refers to the historian's desire to establish the facts of the historical process with a maximum of exactitude. When the historian does his research well we can compare him with Collingwood's well-known detective who wishes to find the murderer of John Doe: he wants to know what actually happened, who did or wrote what, how texts should be interpreted and so on. A number of "auxiliary sciences" (of which modern socio-economic history is the most conspicuous) have been evolved to assist the historian in his attempt to establish the facts. But a historian is essentially more than a "fact-finder" or a detective. Getting to know the facts is only a preliminary phase in the task he sets himself. For his real problem is how to integrate these facts into a consistent historical narrative.[68]

---

[66]*A Study of History*, vol. 1, (2nd ed. London, 1935), 4, quoted by Frank, pp. 11–12. For a real assessment of Mommsen's achievement in a couple of pages, see M. I. Finlay, " 'Progress' in Historiography," *Daedalus*, 1977, 106,3:125–42, at pp. 127–29.

[67]Quote by Frank, 10–11.

[68]F. R. Ankersmit, *Narrative Logic: A Semantic Analysis of the Historian's Language* (The Hague, Boston, London: Martinus Nijhoff, 1983), 8.

While an excessively sharp line should not be drawn between the two aspects of history, Ankersmit explains, facts being "generally only looked for and described within a specific frame of narrative interpretation,"[69] historical practice itself argues against those who deny the distinction altogether. There are in fact historians who are predominantly research historians and historians who are predominantly interpretative historians.[70] Ankersmit concludes that the two activities of research and interpretation or writing have distinct philosophies that account for what they do. To investigate how historical events come to be established is not the province of "narrativist philosophy." That, he says, is "the department of the philosophy of historical research," adding prudently that this "in no way commits narrative philosophy to the view that the historian is free to fabricate historical events."[71] Historical narratives are built up, in short, out of elements borrowed from a repertory of facts or events commonly recognized as historical thanks to the work of historical scholars.

Despite its obvious tenacity, the sharp division—essential to White's argument and to the incommensurability thesis—between historical research and the unstructured historical record on the one hand, and historical narrative on the other, between "facts" and "meaning," "science" and "interpretation," what is found and what is created by a poetic act, is not something simple or obvious. It is not only historically located, as I have tried to suggest, but is itself a philosophical position, as White's best and most astute critics have not failed to observe. White, it has been pointed out, appears to consider the categories of rhetoric that constitute the foundation of his formal analyses of historical texts as themselves purely descriptive, unproblematically given, innocent of any "irrational" act of will or decision, and therefore not subject to critical analysis. Rhetoric, in other words, seems to enjoy a scientific status or a status as knowledge denied to history. The area of decision is thus clearly demarcated from that of knowledge. This is the aspect of *Metahistory* that has come under fire from younger critics, such as Wilda Anderson at Hopkins and two former students of mine, Suzanne Gearhart of U.C.S.D. and David Carroll of U.C. Irvine. All three are more sophisticated philosophically than I and I hesitate to interpret their

---

[69] I made this point myself in "History and Literature: Reproduction or Signification," in R. Canary and H. Kozicki, eds., *The Writing of History: Literary Form and Historical Understanding* (Madison: Univ. of Wisconsin Press, 1978), 3–39, at pp. 31–32.

[70] "There are many historians who have an exclusive interest in historical research: they are concerned with establishing how cities or convents acquired legal or feudal rights, how historical monuments came to be erected, how diplomatic treaties came into being, they study changes in the price of bread or the growth and decline in the population of different areas. . . . On the other hand, there are the historians with a more synthetic turn of mind. . . . They try to integrate the facts found by historical research into large overall views of (parts of) the past. They are concerned not so much with the facts themselves . . . as with what might be the most acceptable representation or synopsis of parts of the past. Their problem is how the history of the past should be narratively written or which narratio proposes the best interpretation of (parts of) the past" (Ankersmit, 8–9).

[71] Ankersmit, 11.

objections. The general thrust of their criticism, if I read them correctly, is that White's absorption of history into literature leaves intact, indeed depends on a more fundamental and traditional distinction—which White himself never questions but which they are eager to challenge—between literary or poetic language and "literal" or scientific language.[72] Because it is believed to be purely referential, the latter is not considered subject to the elaborate formal analyses that the critic makes of the former. "White's typology is only apparently formal," Gearhart contends.

Underlying its fourfold categorization is an opposition between figurative and literal language, and any distinctions between the various forms are ultimately of secondary importance next to this master opposition between language that is essentially formal in nature and language that is essentially referential and, as a result, cannot be subjected to formal analysis.

After all, Gearhart adds, in an extension of an argument that has often been applied to the literary genres, the tropes themselves were not laid up in heaven.[73] Wilda Anderson expressed the objections of all three critics succinctly when she charged that White's metahistorical model depends on a non-discursive and non-historical definition of knowledge.[74]

---

[72]See, for instance, the recent number of *New Literary History*, vol. 17, fall 1985, devoted to "Philosophy of Science and Literary Theory."

[73]Suzanne Gearhart, *The Open Boundary of History and Fiction*, 64. See also David Carroll, *The Subject in Question* (Chicago and London: Univ. of Chicago Press, 1982), ch. 5, and his review of White, "On Troplogy: The Forms of History," in *Diacritics*, 6, no. 3. The argument that genre categories are not universal but language and culture specific has been made forcefully by Michael Glowinski, "Die literarische Gattung und die Probleme der historischen Poetik" (1969), in Aleksandr Flaker and Viktor Zmegac, eds., *Formalismus, Strukturalismus und Geschichte. Zur Literaturtheorie und Methodologie in der Sowjetunion, CCCR, Polen and Jugoslawien* (Kronburg/Taunus: Scriptor Verlag, 1974), 155–85. Thus the terms "fiction," "novel," and "romance" in English allow for the formulation of a theory of prose that would be difficult, according to Glowinski, in Polish, where there is no exact equivalent of the term "fiction" and "romance" is only an archaic designation of the novel. In addition, Glowinski argues, the genres did not arise as objective and disinterested descriptive terms in a poetics whose object is purely scholarly and scientific; instead they serve to express a conception of literature that is constantly being redefined, to enunciate what literature is and is not, what is legitimate in one type of literary expression and illegitimate in another. "Die Literaturwissenschaft schöpft auf diese Weise ihre Gattungsbegriffe aus einer historisch determinierten Praxis, und daher ist eine ihrer Aufgaben die Rekonstruierung der historischen Implikate des Gattungsbegriffes, den man als eine quasi-neutrales Instrument der Deskription benutzt. Der Gattungsbegriff selbst sollte namlich Gegenstand der Forschung, Objekt der Rekonstruktion sein; er ist durch seine Geschichte relevant" (pp. 160–61).

[74]"Dispensing with the Fixed Point," *History and Theory*, 1983, 22:264–77, at p. 276). For another critique of White's rhetoricism, from a totally different standpoint, see Leon Pompa, "Narrative Form, Significance and Historical Knowledge," in David Carr, ed., *La Philosophie de l'histoire et la pratique historienne d'aujourd'hui* (Ottawa: Univ. of Ottawa Press, 1982), 143–57.

V. HISTORY AS PRODUCT AND AS PROCESS

History was perceived by the Romantics as a divine text of which the historian was to serve as the faithful interpreter.[75] Hayden White's view of criticism and interpretation is more modern. For him the interpreter is a creator, a poet in his own right, and he implicitly rejects—as do many contemporary literary critics—the rigid, ultimately theologically grounded distinction between poet and critic, the divine Author of history and the historian as His faithful interpreter. Nevertheless, his view of historiography remains firmly text-oriented, and his emphasis has already been, as he himself puts it, on "the great historical classics," on "historians and philosophers of distinctively classic achievement."[76] But to place the emphasis there is surely to prejudge the issue. History inevitably becomes subject to literary categories of analysis the moment the decision is taken to consider it in its classic texts, that is to say in a select body of timeless, finished verbal structures that can be and have been evacuated from the ongoing processes of which they were once part and kicked upstairs, as it were, to permanent seats in historiography's House of Lords. (In a paper on "Thucydides as 'History' and 'Literature' " Sir Kenneth Dover pointed out recently that it is the very advance of historical techniques that, by emancipating the historian from his dependency on historiography, has isolated the traditional historiographical texts and encouraged us to treat them as "literature," and to focus more on their internal textual structure than on their "external relationship to events."[77]) It is not insignificant that Pierre Bayle, though he is surely one of the founding fathers of modern historiography, rarely figures in the elite company of the "great classics." Bayle's doggedly critical stance, his seemingly deliberate refusal of the whole idea of the "great" or "authoritative" work and his calculated choice of a kind of parasitical strategy of response and commentary (later adopted again by Diderot in many of his best

---

[75] According to Ranke, God "dwells, lives, and can be known in all of history. Every deed attests to him, every moment preaches his name, and most of all the connectedness of all history. This connectedness stands before us like a holy hieroglyph." To decipher the hieroglyph is to serve God as priest and teacher. (Quoted by Leonard Krieger, *Ranke: The Meaning of History* [Chicago: Univ. of Chicago Press, 1977], 361).

[76] "The Historical Text as Literary Artifact," in *The Writing of History*, ed. Canary and Kozicki, 59; *Metahistory*, 4.

[77] *History and Theory*, 1983, 22:54–63, at 59–60. The questions involved in "understanding Thucydides as a whole"—such as, what are his view of human nature and the human predicament, what patterns and cycles are discovered in his story—are indistinguishable, according to Dover, from the questions that "could be asked of *Daphnis and Chloe* or the *Odyssey*."

writings) have resulted in his being almost totally neglected by those whose standard is the monument, the self-enclosed classic text.[78]

From this classic perspective historical narratives are of course incommensurate, and in a sense beyond judgment, as great works of art and music or great heroes and geniuses are supposed to be. They are to be studied, admired, interpreted, and, if possible, imitated. The potentially quite conservative political ethos underlying such a vision of historiography was amusingly acknowledged by Le Roy Ladurie in an article that appeared in *Daedalus* just short of a decade ago. Entitled, ironically perhaps, "Recent Historical Discoveries," it turned out to be a celebration of the *chefs d'oeuvre* of recent French historiography and of the historians themselves as a kind of Meistersinger of Nürnberg. Le Roy Ladurie points out that nearly all of them have dealt "with remote centuries" and suggests why.

These works are concerned with "longue durée" and "longue durée" is no longer a thing of this world. It has been factually expelled from the twentieth century by the acceleration of techniques and of "progress," by uninterpreted, kaleidoscopic social change. It is therefore normal that the masterpieces of contemporary historiography are more concerned with the seventeenth century than with the twentieth. It is deplorable, but it is a fact.

The great historical masterpiece and future classic is thus associated with the values of a world gone by. Le Roy Ladurie is specific: "I have mentioned the word masterpiece: it evokes the supreme achievement of the candidate which qualifies him as a Master in medieval guilds. And, without having to go as far as that, it is true that historical discovery is quite different from what is called by that name in the exact sciences. It is all up to the creator: he leaves behind himself cultural monuments, elaborated out of vast and precious ensembles, the latter having been recovered from the depths of archives." The article ends with a justification of historiography as an elite production of luxury articles that enhance the quality of life in the best French courtly tradition. The French Republic is not so democratic, Le Roy Ladurie reassures his readers, that it is prevented "from continuing to produce good historians, just as it has excellent wines, renowned restaurants, and Camemberts of the finest quality."[79]

Compared with historical texts, the processes by which historical knowledge is produced, established, criticized, and transformed are not well understood and have not been much studied. Perhaps more

---

[78]See the perceptive comment of Weibel on Bayle's enterprise: "il faut déchirer ce corps qu'on a cru parfait et de ses dépouilles méconnaissables construire un objet sans modèle. Le moment du Dictionnaire des fautes est celui de la pure négativité. Après le moment de la séduction—la lecture—, celui de la lacération: et l'on évitera de tomber dans les errements qu'on dénonce, en s'interdisant d'y donner prise, en refusant d'écrire ce corps de mensonge qu'est le livre de savoir" (*Le Savoir et le corps*, 34).

[79]Emmanuel B. Le Roy Ladurie, "Recent Historical 'Discoveries'," *Daedalus*, 1977, 106 (4):155.

attention needs to be paid to the preparation of articles and position papers and the way these are modified in response to the criticisms of colleagues and editors, to professional debate at colloquia and in the scholarly journals, to the vast literature of reviews and the exchanges they quite often provoke, to more or less formally concerted programs of research that have been stimulated by a challenging and innovative narrative.[80]

In several years of service on the editorial board of two university presses, I have been impressed by the effect that the process of review of a manuscript can have on the final published version. Arguments may be changed, chapters shifted around, entire segments dropped and others added, the organization of a work completely redesigned in response to the often surprisingly detailed criticisms of press readers. I am not talking about minor cosmetic changes, or changes intended to communicate the initial argument more effectively, though these are obviously the most frequent. I am talking about truly substantive changes which significantly alter the character of the original manuscript as first submitted. In response to what, I wonder, are such changes made? How is one brought to adopt a perspective one has hitherto rejected, to "see" a configuration that one did not see before, and to substitute this new configuration for an earlier one? Is it pressure, the fear of being isolated professionally, the desire to maintain or acquire professional standing? How, to put the matter in a nutshell, does one change one's mind? Are there *reasons* for doing so, or only causes and motives?

Let me advert for a moment to a humble case of mind-changing from my own experience. For several years I have been studying the society and culture of the Swiss city of Basle in the nineteenth century, having originally been intrigued by the fact that a small, oligarchic, and from most points of view anachronistic city-republic, wedged between France, Germany and the other Swiss cantons, and governed for centuries by a politically conservative elite of some fifty merchant families, served as a sanctuary of intellectual dissidence and a focus of radical speculation in the second half of the nineteenth century. Largely under the influence of Godlmann's work on the seventeeth century French Jansenists—and of Thomas Mann's *Buddenbrooks*—I had devel-

---

[80]After writing the bulk of this paper, I came on Martin Rudwick's stimulating study of a nineteenth-century controversy in geology, *The Great Devonian Controversy: The Shaping of Scientific Knowledge among Gentlemanly Specialists* (Chicago: Univ. of Chicago Press, 1985). Rudwick makes about science itself—interestingly, about the most historical of the sciences—the point I have been trying to make about history. See also Hans Albert, *Treatise on Critical Reason*, 49: "Today's theory of science . . . still displays traits that appear intelligible only within the framework of classical foundationalism. Among them are . . . theoretical monism; . . . the emphasis on the axiomatic method, by which one can develop and justify a privileged theory; and in general an emphasis on the static, the structural, and formal aspects of knowledge, at the expense of its dynamic aspects, which means ignoring developments, conflicts, and the need to choose between alternatives."

oped an elegiac narrative of a business class that was losing its grip and, because of that, found itself in tune with such critics of nineteenth century progressivism as Burckhardt and Bachofen, Nietzsche and Franz Overbeck. This schema was, I now believe, full of flaws. But at the time I did not see them. On the contrary, for reasons I suspect were closely connected with my own personality and my own general feeling of political disillusionment, I was very attached to my schema and had invested it with a good deal of psychic energy. One day, Carl Schorske visited the seminar I was teaching on Basle and Berlin, in order to hear a paper by one of the students. In the course of the discussion that followed Carl, quite innocently, began to draw a picture of the Basle patriciate that was totally at odds with the one I had developed. Where I saw them clinging to old ways of doing business and falling behind economically and commercially, Carl presented them as feisty entrepreneurs with an eagle eye on the main chance. I still recall his referring to their respect for "smarts." I do not think a physical blow would have wounded me more. But soon, because of my respect for Carl (we had taught the course together in a previous year), I began slowly and then with increasing momentum, to look into the commercial habits and practices of the Basle businessmen. I read economic histories, company histories, histories of banking and of the silk and chemical industries, and I gradually came to the conclusion that Carl was right, certainly more right than I had been. A new narrative gradually began to take shape in my mind. No doubt if it is ever written, it will conform to one of the major tropes and will in that sense be subject to metahistory. But neither my ideology nor my esthetic preferences, nor, I am afraid, my basic personality structure, changed much during the period I am talking about. So it does seem to me that it was the discovery of significant new evidence, provoked admittedly by an important psychic interaction with a senior scholar to whom I am very attached, that led me to revise my earlier pattern.

The way historians communicate with each other and criticize each other's work suggests that they do indeed expect their colleagues to be able to recognize the force of contrary arguments and narratives and to adjust their own accordingly—either by developing answers to these arguments or by revising their own. Let me look for a moment at a well known case. The remarkable study of slavery in the Southern States by Robert Fogel and Stanley Engerman, *Time on the Cross*, created a stir not only because it claimed to set a new standard in historical method through the use of quantitative data and analysis, but because it advanced, in addition, a striking thesis about slave society in the antebellum South. For both reasons, it was carefully reviewed in a large number of professional journals—journals of history, Southern history, black history, economic history, sociology, and so on. I have not studied all these reviews, but I have looked at a good many, as well as the important book-length rebuttal by Herbert Gutman, *Slavery and the*

Numbers Game. I don't see how any one who takes the trouble to read even some of this material could fail to be impressed by the very high standard of the discussion. But it is the character of the discussion that I want to emphasize here.

Fogel's and Engerman's thesis, briefly stated, is that, contrary to common opinion, slavery was not an unprofitable system, and that the slaves were anything but lazy, backward, and unteachable. On the contrary, the system worked well, was highly efficient and productive, and provided the South—including the slaves themselves in comparison with the European industrial proletariat—with a far higher standard of living than that enjoyed by most European countries at the time. In addition, the slaves were, on the whole, energetic, disciplined, productive, and capable workers. Because the Afro-American slave, contrary to the arguments of allegedly racist historians, had fully internalized the Protestant work ethic and the mores of Victorian family life, he was able to function in a market economy and society, and though he was in effect confined to the lower levels of the socioeconomic hierarchy, he did strive to rise as high in that hierarchy as he could, thereby contributing to the profits of the enterprise that depended on his labor.

As for their methodology, in beautifully clear and honest prose, Fogel and Engerman outline a position that seems generally positivist (I use this as a descriptive and not as a derogatory term): It is an enormous body of new evidence that has been the source of many of their "discoveries"; the *findings* of the cliometricians should always be distinguished from their attempts, as historians, to interpret them, for these attempts "do not stand on the same level of uncertainty"; social science is incapable of producing the "seamless web" that historians desire and that is woven from all the strands of human behavior studied separately by social scientists—"economic, political, psychological, and cultural"; it provides instead "particular bodies of knowledge," which it is extremely difficult, perhaps impossible to totalize into a unified theory; the links by which restricted areas of relative certainty are joined together to create interpretative configurations are strongly subject to ideological interference.[81] In a passage that in the end is probably not incompatible with the seemingly quite opposite position of Hayden White, Fogel and Engerman explain that "comprehensive ideologies . . . are . . . tempting, because they offer an easy solution to problems of interpretation: they provide the substance needed to cover over the broad and irregular seams of an imperfect historiography and give the impression of a neat seamless web." Though our authors have tried to resist such temptations, they do not claim to have expunged all ideological influences from their book, since in fact "in the main text we attempted to weave these

---

[81]Robert Fogel and Stanley Engerman, *Time on the Cross: The Economics of American Slavery* (Boston: Little, Brown and Company, 1974), 2 vols., vol. 1, Prologue, 8, 10; vol. 2, Appendix A, 34.

new findings into a fairly comprehensive reinterpretation of the nature of the slave economy."[82]

In keeping with their candidly acknowledged positivism, Fogel and Engerman radicalize the common division of the historiographical text into the elegant continuous narrative above stairs and the nitty-gritty scholarship down in the kitchen below by organizing their work in two separate volumes: one to present the thesis or narrative, the finished dish, and the other to display the evidence, the ingredients and processes, which went into producing it. The vigorous disapporval with which many historians reacted to this strategy suggests that the implications of such an extreme division between historical narrative or argument and historical scholarship were sensed and resisted by the profession.[83]

In fact, the criticisms of Fogel and Engerman address *all* aspects of their work, the narrative as well as the "findings." Both are obviously assumed to be subject to discussion, comparison, and evaluation on other than ideological or esthetic grounds. The critics do duly point out both the polemical thrust of *Time on the Cross* and its ideological slant. The book seeks "to vindicate American blacks by defending them against three sorts of defamation"—to wit, that they are inferior racially; or were irreparably degraded and demoralized by the experience of slavery; or lack the kind of initiative and enterprise that other peoples have[84]—and to destroy "myths that turned diligent and efficient workers into lazy loafers and bunglers, that turned love of family into disregard for it, that turned those who struggled for self-improvement in the only way they could into 'Uncle Toms.' "[85] By trying to rehabilitate American blacks in this way, however, Fogel and Engerman are said to be expressing an ideological commitment to "the basic Smithian view that the search for profit via market exchange is a 'natural propensity' of humankind." Their revision of the traditional interpretation of slavery, which represented it as economically irrational behavior, can thus be read as an attempt to remove what was an anomaly from their point of view and thus to reconfirm their vision of the world.[86] As Gutman put it, "Sambo, it turns out, was really a black Horatio Alger, made so by his owner, who was nothing more than a rational profit-maximizer." Thus both "the enslaved and their owners performed as actors and actresses in a drama written, directed, and produced by the 'free market.' "[87]

---

[82]Vol. 2, p. 4.

[83]See, for instance, William J. Wilson and Immanuel Wallerstein in their respective reviews, both in *American Journal of Sociology*, 1976, 81:1192 and 1201.

[84]See W. Letwin in *Journal of Economic Literature*, 1975, 13:60.

[85]Quoted in Herbert Gutman, *Slavery and the Numbers Game*, Preface. According to Fogel and Engerman themselves, "The typical slave field hand was not lazy, inert, and unproductive. On average, he was harder-working and more efficient than his white counterpart" (vol. 1, p. 5).

[86]Immanual Wallerstein in *American Journal of Sociology*, 1976, 81:1206.

[87]Gutman, Preface and p. 1; see also p. 16.

Many of the reviewers—Paul David, Martin Duberman, Peter Temin, Herbert Gutman himself—make the point that Fogel and Engerman "save" the blacks only by integrating them into the mainstream of American culture.[88] Gutman reinforces this point when he proposes that *Time on the Cross* can be regarded as a belated contribution to the consensus historiography of the 1950s, in which the motif of conflict in American history was underplayed while greater emphasis was placed on the way political adversaries were often bound together by shared values. With Fogel and Engerman, Gutman argues, the synthesis is now stretched to include even the most recalcitrant element, the Afro-American.

The exposure of ideologically loaded assumptions and political and economic values is not, however, equivalent to a denunciation of such assumptions and values as totally irrational. On the contrary, it is an invitation to those who have them, above all to Fogel and Engerman themselves of course, to become cognizant of the fact that they do, and to justify them or reconsider them. It is only because assumptions are themselves, in some measure, at least discussable—that is, in the sense that a rational person ought to be accountable for his assumptions and values and to be able to defend them, if not to *demonstrate* them—that it is worth exposing them in public discussion of issues in the first place. But the debate over *Time on the Cross* extended beyond basic ideological commitments. Reading the criticisms one has the sense that a great deal of what historians do is in fact subject to fairly stringent criteria of evidence and reasoning, criteria acknowledged by the entire profession, whether it is being practised in the West or in the East, by Marxists or conservatives, in rich countries or in poor ones. Gutman, for instance, proposes to ask "a variety of questions about the evidence and arguments in T/C. . . . These questions," he claims

are appropriate to all historical works and to all sorts of historical evidence. Have the authors asked the right questions? Have the questions asked been answered properly? Have the right sources been used? Have the sources used been properly studied? Are there conceptual errors in the use of the sources? Are there errors in what quantitative historians call 'executional computations'? Has the work of other historians on similar subjects been properly used? Have the arguments of other historians on similar subjects been properly summarized? How do the new findings measure against the published findings of other

---

[88]"To have been a hard-working, responsible slave is to have been part of the 'positive' development of black culture, to have contributed to the saga of black 'achievement.' F and E berate Stampp and Elkins for having portrayed slaves who lied, stole, feigned illness, acted childishly, shirked their duties—as if these traits unnecessarily reflect badly on black personality and culture. But they do so only in the context of a Calvinist work ethic and the Boy Scout pledge to loyalty and cheerfulness. . . . Themselves equating 'efficiency' with 'achievement' and eager to award blacks their credentials in middle-class white culture, F and E reject any suggestion that a slave's refusal to become an effective cog in the plantation machine can itself be seen as an achievement, a testimony to black ingenuity, resistance, pride" (Duberman, quoted by Gutman, 165, note 229).

historians and against other sources not examined by the author? What is the relationship between 'hard' empirical findings and speculative inferences and estimates?[89]

The criteria here are admittedly criteria of rightness and propriety, that is to say that they refer to the rules and codes established by a community, but that does not place them beyond rational discussion.

Gutman's debate with Fogel and Engerman ranges in fact from detailed questioning of their sources and their utilization of them to a much broader questioning of the model of socialization that they use in their study of the slave. I shall give only one example of innumerable detailed criticisms. It concerns the computation of the frequency of slave whippings. Fogel and Engerman argue that whipping was not a crucial part of the slave system.

What planters wanted was not sullen and discontented slaves who did just enough to keep from getting whipped. They wanted devoted, hard-working, responsible slaves who identified their fortunes with the fortunes of their masters. Planters sought to imbue slaves with a "Protestant" work ethic and to transform that ethic from a state of mind into a high level of production . . . Such an attitude could not be beaten into slaves. It had to be elicited.[90]

Using the diary kept by one plantation owner, well known for his belief that to spare the rod was to spoil the slave, they compute "an average of 0.7 whippings per hand per year." While on the face of it, this figure is accurate enough, it is not, Gutman argues, the significant average. "The wrong question has been asked." One could just as well have come up with a figure of 0.013 whippings per hand per week, which seems even more insignificant. Gutman then cites the case of lynchings at the end of the last century. It is known that "on average" 127 blacks were lynched every year between 1889 and 1899. "It is useful," he asks, "to learn that 'the record shows an average of 0.0003 lynchings per black per year, so that about 99.9997 percent of blacks were not lynched in 1889?" The essential statistic, in short, is not the average number of whippings per hand. Such a figure does not measure the utility of the whip as an instrument of social and economic discipline. Is it much more relevant to know that "slave men and women were whipped frequently enough—whatever the size of the unit of ownership—to reveal to them (and to us) that whipping regularly served as a negative instrument of labor discipline."[91]

Gutman questions not only the way Fogel and Engerman used the material in Barrow's diary but their reading of it, and he argues plausibly that in fact the figure of 200 slaves on Barrow's plantation, the figure on which Fogel and Engerman based their computations, was arrived at by a series of erroneous assumptions and reasonings. 129 turns out to be a

---

[89]Gutman, 11–12.
[90]*Time on the Cross,* vol. 1, p. 147.
[91]Gutman, 19–20.

more likely figure, so that their averages are not only irrelevant, they are probably wrong. By throwing doubt on the data on which Fogel and Engerman constructed their narrative, Gutman threatens the validity of that narrative as history. No internal coherency or plausibility can save it. It may be a good story, and it may make a good point about life or human nature or whatever, but it will not pass as what it purports to be—a history of slavery in the antebellum South—unless Fogel and Engerman somehow reestablish the validity of the elements from which they have constructed it.

It is hard to tell at what point the weakness of the component parts disqualifies a historical narrative irremediably. Some arguments or narratives can apparently stand a good deal of buffeting and erosion without collapsing. The various elements historical narratives are built up out of are obviously of varying structural importance.

In his review of *Time on the Cross* Gutman recognizes a distinction between technical flaws and structural weakness. "Errors mar F+E's basic findings on slave sales, slave marriage and family ties, slave sexual behavior, slave punishments and rewards, and the urban and rural slave occupational structure. But if the estimates were more accurate and the quantitative data examined more soundly, the model meant to explain slave beliefs and behavior would still be inadequate."[92] For one thing, the narrow, exclusively economic explanations given by Fogel and Engerman may be necessary but they are not sufficient.[93] In addition, Gutman claims, in order to explain the beliefs and behavior of the slaves, Fogel and Engerman have used "an analytic model as old as the pioneering work of U. B. Phillips, one that had its most sophisticated expression in the writings of Stanley Elkins"—two scholars who are severely criticized in *Time on the Cross*.

Neither Elkins nor F+E, of course, share in any way the mistaken racial assumptions that irreparably damage all of Phillips' writings. They nevertheless share a conceptual model with the founding father;. . . . a model which views slave belief and behavior as little more than one or another response to planter-sponsored stimuli. . . . More needs to be asked than the question Phillips posed. Simply changing the factors in the Phillips equation and, of course, rejecting his racial assumptions does not necessarily make for more truthful answers or better social history. We need to know in close detail what enslavement did to Africans and then to their Afro-American descendants. But we shall never comprehend slave belief and behavior by just asking that question. We need also to ask what Africans and their Afro-American descendants did as slaves. That is a very different question and the answers to it are not

---

[92]Gutman, 167.

[93]Gutman refers to criticism by David Fischer and Harold Woodman, both of whom denounce as a "reductive fallacy" the confusion of "a causal component, without which an event will not occur, with all other causal components which are all required in order to make it occur"—a confusion often found "in causal explanations which are constructed like a single chain and stretched taught across a vast chasm of complexity" (Gutman, 167). This criticism seems to imply acceptance of some version of Hempel's "covering law" theory.

the mirror-image of what owners did to slaves. The novelist Ralph Ellison put it well in criticizing the main thrust that has informed the writing of much Afro-American history. His criticisms apply just as well to T/C. "Can a people . . . live and develop over three hundred years by simply reacting?" asked Ellison. "Are American negroes simply the creation of white men, or have they at least helped create themselves out of what they found around them?"[94]

Fogel and Engerman tell a very good story in my view. But Gutman's criticism of it is neither esthetic nor by any means exclusively ideological: he believes that both the empirical scholarship and the explanatory model used are faulty and that another story which was not so flawed would be *better*. This kind of criticism is such common practice among historians that it seems hardly necessary to have gone to such lengths to illustrate it. But if we want to understand what historians are doing, we must recognize the preponderant place such work occupies in their overall activity. Gutman's own book is also a work of history, although it is exclusively a rebuttal of Fogel and Engerman and is not itself a narrative. In this respect, it follows a model that was established at the very beginnings of modern historiography. Bayle's writing, as I have already mentioned, was almost always critical, whether we think of his reviews in the *Nouvelles de la République des Lettres*, of the articles in the *Dictionnaire historique et critique*, or of more extended works like the *Critique de l'histoire de Maimbourg*. In this respect, modern historiography may well stand closer, as has sometimes been observed, to the genre of literary criticism than to that of the epic or the novel.

Historians do apparently believe that there are procedures of verification and criteria for judging between different hypotheses and different narratives. They do constantly attend to the research of their colleagues, to see whether it corroborates their own findings or not, and they build upon the results of the work of others both to modify hypotheses that they currently hold and to develop new hypotheses and new programs of research. Le Roy Ladurie, for instance, used at one time to take special pleasure in referring to theses and articles still unpublished that bore upon his own work and were in fact largely stimulated by it, as though to underscore the collective character of historical research. (That was in the days when the leading French historians were still optimistic about the scientific vocation of their discipline.) Fogel and Engerman also acknowledge receiving help in the writing of *Time on the Cross* from innumerable colleagues, including many whom they criticize (like Stampp) or were subsequently criticized by (like Gutman). The work of the modern historian is no solitary activity in which a person of imagination constructs a narrative to convey a certain vision and certain values, while paying a kind of ritual respect to the "conventions" of historiographical writing (such as checking on evidence, and the like). That *might* have been the case in the

---

[94]Gutman, 170.

eighteenth century—though even then pioneers like Malthus, Playfair, and Eden were outlining a different *kind* of historiography, one more indebted to Montesquieu and to Adam Smith than to Livy and Tacitus[95]—or in the early nineteenth century, especially in France, when history was not yet organized professionally and was still, as Monod remarked in the Preface to the first number of the *Revue historique*, the product of individuals of genius. Clio and Calliope then still apparently retained something of that union with each other and with all their sister Muses which was supposedly their original condition, the division of labor among them and the attribution of specific and distinct functions to the individual Muses being a late invention, we are told, of rationalizing Hellenistic scholars and commentators.[96]

We do well to remember that our present disciplinary boundaries were no more laid up in heaven than the boundaries between living species. At the same time, there cannot really be any question of rejecting the differentiations and the specialized techniques and procedures of analysis that have been elaborated through the ages in order to make human knowledge and action in the world more effective. Such an ambition, as Stanley Fish pointed out in a recent defense of professionalism,[97] is comparable to the Romantic desire to leap back from language into some impossible immediacy of expression, untouched by collectively developed differentiations, definitions, and procedures. The differentiation of history from fiction, and of both from myth,[98] and, more recently, the professionalization of history have

---

[95]See Frederick Morton Eden, *The State of the Poor, or An History of the Labouring Classes in England from the Conquest to the Present Period, in which are particularly considered their Domestic Economy with respect to Dress, Fuel and Habitation [. . .] with a Large Appendix containing a comparative and chronological table of the prices of Labour, of Provisions, and of other Commodities, an account of the poor in Scotland, and many original documents on subjects of National Importance* (London: 1797); William Playfair, *The Commercial and Political Atlas* (London, 1786); the same author's *Inquiry into the Permanent Causes of the Decline and Fall of Powerful and Wealthy Nations* (London, 1805); and Thomas Robert Malthus, *An Essay on Population* (London, 1798). Malthus's text is worth quoting. The reason why certain underlying regularities in human history have not been noted, he claims, is that "The histories of mankind which we possess are, in general, histories only of the highe. classes. We have not many accounts that can be depended on of the manners and customs of that part of mankind where these retrograde and progressive movements [of population] chiefly take place. A satisfactory history of this kind, of one people and of one period, would require the constant and minute attention of many observing minds in local and general remarks on the state of the lower classes of society, and the causes that influence it; and to draw accurate inferences upon this subject, a succession of such historians for some centuries would be necessary. This branch of statistical knowledge has, of late years, been attended to in some countries, and we may promise ourselves a clearer insight into the internal structure of human society from the progress of these inquiries . . ." (Everyman Liberty, ed., London: J. M. Dent, 1914, vol. 1, pp. 16–17).

[96]See Adolf Trendelenburg, *Der Musenchor: Relief einer Marmorbasis aus Halikarnass* (Berlin, 1876 [Winckelmannfest der Archäologischen Gesellschaft, Program 36]), 13; Max Wegner, *Die Musensarkophage* (Berlin: Mann, 1966), 98–99, 108. Also Elisabeth Schröter, *Die Ikonographie des Themas Parnassus vor Raphael* (Hildescheim and New York: Georg Olms, 1977), 176–77.

[97]"Anti-Professionalism," in *New Literary History*, 1985, 17:89–108.

[98]Louis Mink himself emphasized the crucial importance of these distinctions: see the

resulted in the elaboration of highly refined and carefully monitored methods of investigation. Every affirmation is now much more subject to scrutiny and control than in the age of traditional histories or the heady time of charismatic Romantic historiography. Fraud and shoddy application of research techniques are notions that are as real to historians as to physicists or chemists and are perceived by both groups as threats to the viability of their discipline. The writing and rewriting of histories is no longer, for the most part, a matter of adapting a fairly well established story line to a new language or rhetoric, or even of interpreting it in a new way, "emplotting" it differently, and thus expressing through it an ideology or a vision of the world. It is not just a matter of "decoding" and "recording" a set of events that the historian simply comes on as given and that have been established as a result of a totally autonomous operation,[99] even if such strategies are still followed in some branches of historiography.[100] It is also, perhaps primarily, a matter of discovering and responding to new professional and scientific exigencies, new kinds of questions, new evidence that either provokes new questions or has been turned up as a result of the invention of new questions, participating in critical exchanges with colleagues.

In this respect history, it seems, is not radically different from other kinds of scientific inquiry. In his study of *The Great Devonian Controversy* Martin Rudwick observes that historians of science have focused too much on the work of single individuals and have neglected the "complex web of social and cognitive interactions that bind even the most distinguished or reclusive scientist into his or her immediate network of colleagues, in collaboration, or rivalry, or both." The Devonian controversy "displays the processes of scientific knowledge making as eluctably and intrinsically social in character, not (or not primarily) in the sense of the pressures of the wider social world, but in the sense of intense social interaction among a small group of participants. But it also shows," Rudwick adds, "that the knowledge produced through this interaction is not 'merely' a social construction, and that the concrete natural world does have an identifiable input, constraining though not determining the eventual outcome of the research." In the particular instance of scientific "knowledge making" that he studied, Rudwick claims that

---

conclusion of his essay in Canary and Kozicki, *The Writing of History,* 148–49.

[99]Hayden White, in Canary and Kozicki, *The Writing of History,* 59.

[100]See, for instance, Peter Paret's review of Nigel Nicholson's *Napoleon 1812.* (*New York Times,* January 26, 1986). Paret distinguishes here between "academic" or professional and "nonacademic" or popular history: "A professional historian writing a new book on 1812 would need to justify his project by presenting new documentation or revising accepted interpretations. Mr. Nicolson, who is not an academic, neither needs nor wants to add to our knowledge of the invasion. He is content to give a coherent, well-informed account of an episode that clearly has long fascinated him and that he has studied seriously. The value of such nonacademic history is not to be underestimated."

while the "Devonian" case was argued out in courtrooms such as the Geological Society by the persuasive advocacy of major participants [and Rudwick emphasizes the rhetorical element in *all* scientific reasoning] other competent geologists, who in effect constituted the jury in the case, were not swayed into a consensual verdict only by the rhetorical skills of those presenting one side of the case, still less by being bribed out of court by promises of advancement in their careers [. . . .] What swayed them was the combination of rhetoric and evidence, persuasive argument and *pièces justificatives*. Like a court of law, and unlike, for example, a debating society, the geologists were concerned not with evaluating rhetorical performance but with reaching a justifiable conclusion about concrete past events in a real world.

Rudwick adds this important comment: "Significantly, both sides were obliged to shift their position during the protracted hearing, in response to the other's arguments and evidence, and the final verdict was based on a case that neither side had anticipated when the hearing began."[101]

The adversarial method that characterizes the search for the best solution in our courts of law, our representative institutions, and—no less significantly, in modern times at least—our economic system, is also characteristic of scholarship as we practice it today. Out of the clash of conflicting arguments and theories, it is believed, the best solution will emerge and will ultimately be recognized by all. Such was the optimistic liberal conviction that was expressed in 1830 by Henry Thomas De la Beche, a geologist attached to the Ordnance Survey of Great Britain and a director of the Museum of Economic Geology in London:

That much good ensues, and that the science is greatly advanced by the collision of various theories, cannot be doubted. Each party is anxious to support opinions by fact. Thus, new countries are explored, and old districts re-examined; facts come to light that do not suit either party; new theories spring up, and in the end, a greater insight into the real structure of the earth's surface is obtained.[102]

One could imagine a more co-operative system in which conceding another's point of view was easier. But perhaps arguments would be less keen and less productive in such a system, as well as less strictly and effectively monitored.[103] At any rate, challenging accepted views, changing one's mind, and adjusting one's arguments appear to be essential features of any system in which human beings argue with each other and seek, out of that argument, to arrive at "the truth," at what in legal parlance is "beyond reasonable doubt."[104]

---

[101]*The Great Devonian Controversy* (Chicago and London: Univ. of Chicago Press, 1985), 6, 456.

[102]Quoted by Martin Rudwick as the epigraph to *The Great Devonian Controversy*.

[103]On the value of "sticking to a theory as long as possible," see Lakatos, *Methodology of Scientific Research Programmes*, 89, 92, 118. According to Lakatos, "Purely negative, destructive criticism, like 'refutation' or demonstration of an inconsistency does not eliminate a programme. Criticism of a programme is a long and often frustrating process and one must treat budding programmes leniently" (p. 92).

[104]See Stephen Toulmin, *Human Understanding*, (Princeton: Princeton Univ. Press,

Modern historiography, in sum, is a professionalized and regulated activity in which no individual can any longer imagine that he or she works alone or enjoys a special relation to the past. In this respect it differs from Neoclassical or Romantic historiography. On the whole, I believe this state of affairs is a good one. Every historian is now answerable for his statements and his stories to well informed, critical peers, like a statesman in some ideal democracy. Prophetic utterances and the manipulation of feelings and opinions are disapproved, and failure to observe established procedures results in loss of credibility.

It is sometimes said, however, that professionalism resembles a democracy less than it resembles a bureaucracy, that it encourages routinized scholarship, fear of intellectual independence and risk-taking, and even the avoidance of criticism itself. A conservative desire to maintain the "system" replaces the adventurous spirit of the pioneers of the discipline. This is an important objection. The chief advantage of professionalism is that it ensures responsibility and accountability, opens everything up to critical inspection, and promotes continuity and commonality of research and knowledge, among specialists at least. That advantage would be dearly purchased if it involved the repression of initiative and imagination. A deadening uniformity might even lead to a recrudescence of prophetism, as the imagination, excluded from the channels of professional discourse, assumes an aggressively anti-critical rather than simply non-critical or pre-critical stance. It is essential, therefore, that space be reserved within any professional system for speculation and experiment, even for ideas that challenge the most widely shared and well established doctrines, and that a fair and genuine hearing be granted to even those views that put in question the efficacy of the very programs by which competing arguments are rationally sifted and compared—that the system, in short, allow and provide for its own renewal. It is probably not possible to ensure the preservation of such free spaces in a profession, or in a society, by legislation alone: much will depend on the good will, vigilance, and commitment to freedom of the members of the profession themselves.

1972), p. 388; and Imre Lakatos, *Methodology of Scientific Research Programs*, 117. Lakatos uses a legal analogy to make the point that in nearly all theories of scientific methodology it is an act of reasoned judgment that determines which theories are to be preferred, which abandoned, and at what point, and that also ensures the right of appeal: "The Duhemian conventionalist needs common sense to decide when a theoretical framework has become sufficiently cumbersome to be replaced by a 'simpler' one. The Popperian falsificationist needs common sense to decide when a basic statement is to be 'accepted' [. . .]. But neither Duhem nor Popper gives a blank cheque to 'common sense.' They give very definite guidance. The Duhemian judge directs the jury of common sense to agree on comparative simplicity; the Popperian judge directs the jury to look our primarily for, and agree upon, accepted basic statements which clash with accepted theories. My judge directs the jury to agree on appraisals of progressive and degenerating research programmes. [. . .] Although it is important to reach agreement on such verdicts, there must also be the possibility of appeal. In such appeals inarticulated common sense is questioned, articulated and criticized. (The criticism may even turn from a criticism of law interpretation into a criticism of the law itself.)"

And as civil liberty, in social life, permits the expression of ideas and opinions that the majority judges "outrageous" and that command only a limited audience, professional freedom should and usually does leave room for unorthodox or "crackpot" ideas that are not widely attended to. It may be objected that the marginalization of the unorthodox is the particular form of repression by which an ostensibly liberal profession ensures that its dominant beliefs will not be challenged. As long as liberty of expression is guaranteed, however, there seems no reason to doubt that an unorthodox idea has as good a change of getting a hearing as it is possible to provide. It is up to the proponents of the idea to make a good case for it and it is up to the practitioners of history to maintain an open mind in judging whether it is worth entertaining or pursuing.[105] There is no reason to believe that the commitment of the professionals to the received opinion of the profession is greater than their commitment to the intellectual discipline they pursue. Challenges to received opinion have been made in the past, and many new and disturbing ideas have overcome the objections they provoked to become, in turn, widely accepted. It could even be argued, as we shall see, that the possibility of making such challenges and of achieving changes in accepted opinion is what defines a discipline as rational.

The question that the critics of professionalism are asking may be a more far-reaching one. They may be asking whether professionally regulated languages of communication are capable of change; whether the professionals can recognize the merit and interest of ideas, concepts, categories that challenge their institutionalized discourse and require a considerable, even a wholesale revision of it; and whether such change can occur for "reasons," that is, according to the norms of the system or discipline itself. For if "reasons" can only be defined in terms of the system they ultimately sustain, how can they ever effectively challenge it? Such questions are by no means restricted to the historical profession. They directly concern any science, as we have learned from scholars like Popper and Kuhn, Lakatos and Feyerabend. The commensurability of historical narratives is a more restricted problem, however, which can probably be answered *within* the terms of a given intellectual practice. Comparing two narratives of the French Revolution is more

---

[105]I think of the *Journal of Historical Review,* which sits demurely on a shelf quite close to the *Journal of Modern History* in Firestone Library at Princeton. The *Journal of Historical Review* is set up to argue in issue after issue that the holocaust is a myth, a fabrication of the Jews, a literary, not a historical reality. Though it has no doubt been admitted to the library in the name of freedom of thought and expression, it is doubtful that it commands any serious attention. Does this mean that it has not received a fair hearing and that its presence in the library is the mark of a purely formal freedom? I do not think so. In the debates about the holocaust, arguments which are by no means flattering to the victims and which disturb some deeply held convictions have been seriously considered. If those of the *Journal of Historical Review* have not succceeded in getting themselves taken seriously, it is because the journal has not convinced the professionals that it subscribes to the basic rules of the profession. Its almost parodic execution of the external gestures of scholarship only underlines this difference in essentials.

like comparing two theories of matter than comparing the modern "scientific" understanding of nature with the knowledge of nature and the practices used to affect it in another culture. There may be, in other words, indeed there are, different conceptions of history and different practices of historiography from those that have established themselves in the modern West. The question is only whether, within our present professional framework, it makes sense to compare different historical narratives and whether there are rational grounds for preferring one to another.

VI. THE RATIONALITY OF HISTORICAL ARGUMENT

The problem we have now returned to is the one I defined earlier as "changing one's mind" or, on a larger scale, changing a story or a professional paradigm within a context in which truth, in the sense of some correspondence between ideas and reality, or propositions and facts, can never be demonstrated. What are the processes by which that change occurs? Is there any rationality about them or are they arbitrary, a matter of fashion or desire or ideology? As Ian Hacking put it not long ago in a review of some essays by Thomas Kuhn, if the structure of our science, whether physical or social, "is ultimately a human creation and we populate a world which is partly an artifact of human discourse, then it is less than clear what makes theory-choice 'objective.' "[106] If "objective" here means ontologically founded, then I am not sure how it can be demonstrated that theory-choice is objective. If, however, "objective" has a more modest meaning—something like rationally justifiable or defensible, not arbitrary, open to criticism—then there may be some hope for objectivity.

To find a middle ground between absolute truth (metaphysical and logical truth based on correspondence with "reality" or internal coherency and systematicity) and irrational arbitrariness or "decisionism," as the German philosophers say (truth as a product and instrument of will, desire, interest), to mark out, in other words, a territory appropriate to what human beings can realistically achieve in the matter of scientific objectivity seems to be the principal aim of Stephen Toulmin's *Human Understanding*, the first volume of which (the only one to appear so far) is an extended argument, largely directed against Collingwood and Kuhn, in favor of the continuity of the practice of science as a discipline and of the reality of communication among different theoretical positions.[107] Toulmin holds that we "can no longer afford to assume"—

---

[106]Review of Kuhn, *The Essential Tension*, in *History and Theory*, 1979, 18:222–36.

[107]In fact, Hacking argues that "revolution of a Kuhnian sort is normal, [. . .] only one half of normality, while the other half is what Kuhn chose to designate as normal," for "crises come [. . .] as an integral part of self-conscious normal science, all of which operates within a framework of unstated conditions of what it is possible to say." There is thus a tension, according to Hacking, between Kuhn the philosopher who writes of "incommensurable conceptual schemes" and Kuhn the historian "who provides an internal history of problems and their solutions" (Hacking, op cit.). See also Toulmin on this, *Human Understanding* p. 105 et passim. Stanley Fish's criticism of Toulmin in the article referred to above (note 99) is consistent with Fish's position. I have great sympathy with Toulmin's attempt, which Fish characterizes as ultimately contradictory, to mark out

as both absolutists like Frege and relativists like Collingwood allegedly do—that our rational procedures, if they are to be impartial, must "find a guarantee in *unchanging principles* mandatory on all rational thinkers." We should reject, in his view, "the commitment to logical systematicity which makes absolutism and relativism appear the only alternatives available." Toulmin suggests that we redirect our attention instead from arguments or explanations considered as formal structures in themselves to "explanatory activities and procedures . . . other than those which involve appeal to formal, demonstrative arguments."

A scientific practice, as he proposes we understand it, is "transmitted from one generation of scientists to the next by a process of *enculturation,* [. . .] an apprenticeship, by which certain explanatory skills are transferred, with or without modification, from the senior generation to the junior." The rationality of a science, consequently, is not embodied in the specific intellectual doctrines that are adopted at any given time by individual practitioners or by a professional group as a whole. It is related rather to *"the conditions on which, and the manner in which,* [an individual or a group] *is prepared to criticize and change those doctrines as time goes on,"* to the procedures that allow for discovery and conceptual change through time. The specific character of scientific doctrines, as opposed to ideologies or religious beliefs, for instance, is that they are "historically developing 'rational enterprises' [. . .] committed to their own self-transformation."[108] The ability to change one's mind *for good reasons* thus becomes for Toulmin the very criterion of rationality.

We judge the rationality of a man's conduct by considering, not how he habitually behaves, but rather how far he modifies his behavior in new and unfamiliar situations, and it is arguable that the rationality of intellectual performances should be judged, correspondingly, not by the internal consistency of a man's habitual concepts and beliefs, but rather by the manner in which he modifies this intellectual position in the face of new and unforeseen experiences.[109]

---

a sphere of "reasons" as well as a sphere of "causes" and a set of "disciplinary" criteria distinct from purely "professional" criteria. Essentially, Toulmin is trying to define a role for reason in history. It is obviously simpler and more consistent to deny that it has one and to subordinate one category to another: either history is ultimately rational, or reason is ultimately itself historical.

[108]Toulmin, 51, 157, 84, 165. Cf. Hans Albert, *Treatise on Critical Reason,* 62: "Science progresses neither by the derivation of certain truths from self-evident intuitions with the aid of deductive processes [classical rationalism—L. G.], nor through the derivation of such truths from self-evident perceptions using inductive processes [classical empiricism—L. G.]: it advances, rather through speculation and rational argumentation, through construction and criticism."

[109]Toulmin, p. 486. See also the epigraph to the book: "A man demonstrates his rationality, not by a commitment to fixed ideas, stereotyped procedures, or immutable concepts, but by the manner in which, and the occasions on which, he changes those ideas, procedures, and concepts." In an interesting application of Toulmin and others to the problems of literary scholarship, Jürgen Klein argues that defining scientific and scholarly rationality—as he himself proposes—"nicht als kognitiver Prozess, sondern als intellektueller Prozedur" will spell the end of the reign of hermeneutic. (Jürgen Klein,

The distinction between rationality and irrationality, in other words, is a methodical, not a foundational distinction, and as such it must have reference to practice and find its proper place *within* the sphere of evaluation and decision.

Although history may still be only a "would-be discipline" in Toulmin's terms,[110] it does seem to be more organized both theoretically and institutionally than it was even a hundred years ago. Techniques of discovery and analysis are constantly being expanded and refined, and many of the changes win acceptance regardless of ideological allegiances. "Historical judgments," one contemporary historian has said, whether they take the form of narratives or typological constructs, "are *intersubjectively understandable*. [They] are also *intersubjectively verifiable*." Thus it is possible to formulate a number of questions that all historians will ask as part of the verification of a given hypothesis, such as "(a) to what extent relevant sources have been utilized and the present state of research has been taken into consideration, (b) how close these historical judgments have come to reaching an optimum plausible integration of all available historical data, and (c) how logically rigorous the explanatory models underlying them are, that is, how consistent those models are and whether they are free of self-contradictions." If these conditions are met—or come reasonably close to being met—the system of historical judgments in question will be considered not only "intersubjectively plausible but also 'correct,' though only in terms both of its presuppositions and its methodological approach."

These standards of historical judgment, Wolfgang Mommsen claims, constitute a framework within which it is possible to speak of a "progress" of scholarly knowledge. To the degree that historians subscribe to the standards, they create the possibility of free exchange of information and sustained critical evaluation of historical judgments across social and ideological lines. In the course of time, a body of historical knowledge about certain subjects can be expected to arise that is well established and universally accepted.[111]

Theoriengeschichte als Wissenschaftskritik: Zur Genesis der literaturwissenschaftlichen Grundlagenkrise in Deutschland [Haustein: Forum Academicum, 1980], 86–89). On the longstanding conservative political implications of hermeneutical thinking and the "dubiousness of ontologizing the textual model in hermeneutic philosophy" (an operation unquestioningly accepted by many humanists these days), see Hans Albert, *Treatise on Critical Reason*, 165–98.

[110]"Diffuse" or "would-be disciplines" are defined by Toulmin as those in which "the fundamental explanatory task has not been clearly defined," so that "neither an agreed set of disciplinary methods nor a common forum of disciplinary debate [has been] able to establish its authority" (*Human Understanding*, p. 389).

[111]Wolfgang Mommsen, in *History and Theory*, Beiheft 17, 1978, p. 33. See also the more cautious and nuanced view of Chaim Perelman, "Objectivité et intelligibilité dans la connaissance historique" [1963], in his *Le Champ de l'argumentation* (Brussels: Presses universitaires de Bruxelles, 1970): "La connaissance historique, produit de notre culture— elle-même produit de notre passé—doit assumer, tout comme la philosophie, la perte de confiance, qui caractérise notre époque, dans des critères absolus et infaillibles, en matière d'intelligibilité. . . . S'il est légitime de rechercher, en histoire, des critères sur lesquels

Essentially, Mommsen's idea of how historical knowledge "progress-es" is a version of what Lakatos has defined as "dogmatic falsificationism"—the view that, though all theories are conjectural and science cannot *prove* any, it can *disprove* them, because there is an absolutely firm empirical basis of facts that can be used to disprove theories. Lakatos: "According to the logic of dogmatic falsifications, science grows by repeated overthrow of theories with the help of hard facts."[112] Mommsen:

> The historian's sources alone will define clear limits for the application of whatever explanatory model he chooses, limits he cannot violate with impunity. For every possible interpretation of an historical problem there are always some recalcitrant facts that can be integrated into the interpretation only by modifying or expanding the chosen explanatory model. Worse yet, these facts may prove the model inadequate or even totally untenable.[113]

Mommsen thus seems to bring us back again to that demarcation between theory and experiment, historical narrative and historical research, about which I have already expressed my uneasiness. It also leaves unanswered the question: At what point does empirical counter-evidence refute a theory or invalidate a narrative? How much and what kind of evidence is required? Discussing the incommensurability thesis earlier, I noted that Hayden White considers the great historical narra-tives to be impervious to empirical data on the ground that their truth lies in their systematicity or internal coherence (which is not affected by the truth or otherwise of their individual referential statements) and in the vision of the world that—like great works of literature—they propose or stimulate in successive generations of readers. But even among those who, like myself, would wish to place historical narrative closer to scientific theories than White would allow, only a few diehards would want to contest nowadays that the relation of theory to "facts" or "observations" is by no means a simple one. A theory does not collapse simply because a number of "facts" have been discovered to be inconsistent with it. Indeed, Imre Lakatos argues the opposite. A good theory is one that is so designed that its core is virtually invulnerable to such observational refutation: "Purely negative, destructive criticism, like 'refutation' or demonstration of an 'inconsistency' does not elimi-nate a research programme. Criticism of a programme is a long and often frustrating process."[114] Falsification, in other words, is not the

---

tous les historiens s'accorderaient, il est vain de limiter l'histoire, sous prétexte d'objectivité, à ce qui peut être connu grâce à de pareils critères. Mais dans la mesure même où l'historien reconnaît l'imperfection de son oeuvre, il ne peut se refuser au dialogue avec d'autres historiens, en espérant que ce dialogue lui permettra de présenter un récit moins arbitraire, c'est-à-dire plus complet, plus intelligible et plus impartial du passé" (p. 371).

[112]Lakatos, 12–13.
[113]Mommsen, 33.
[114]Lakatos, 92.

straightforward business some Popperians might have thought. Likewise, the fact that a hypothesis or program is discovered to rest on some questionable observations or data does not immediately disqualify it. The merits and fruitfulness of a hypothesis are not strictly dependent on the observations and data that may have given rise to it or been used initially to support it.

The same may perhaps be said, *mutatis mutandis,* of a historical hypothesis or argument. A case in point is the book on the relation between the Nazi party and the German industrialists in the 1930s by my former colleague at Princeton, David Abraham. David's book was attacked by a distinguished professor at Yale on the grounds of inaccuracies and, as he charged, falsifications of evidence in the argument. (Something, incidentally, that is as threatening to scientists as to historians.) But David's defenders, while acknowledging the shoddiness of his technical procedure in places and admitting the possibility of tendenciousness in certain allegedly inaccurate quotations from the archival materials, held that in the end the criticisms concerned relatively minor points and did not drastically affect the structure or validity of the book's argument. The criticisms thus damaged David's authority and reputation as a historian far more seriously than his thesis itself.

To acknowledge that we do not well understand the relation between evidence and the theories or narratives it is used to sustain or refute does not require us to revert to Hayden White's view that the facts are simply not relevant to the validity of the narrative, and that our grounds for preferring one narrative to another or for changing our minds are esthetic preference or ideological commitment. It does require that we address the problem honestly. If we believe there are rational grounds for preferring one theory or hypothesis or narrative to another and if we hold that the fact that a hypothesis or theory or narrative is potentially capable of being rejected in favor of another "better" one—in other words, its vulnerability to rational criticism—is the very condition of its being considered knowledge in the first place (so that knowledge does not imply certainty; on the contrary, it implies openness to criticism), then it is a matter of great importance to try to understand how and in what conditions a theory does come to seem more or less plausible. Is it its capacity or incapacity to account for accepted facts—in Popperian terms, conflict between accepted theories and accepted basic statements? Or does it involve, as Lakatos would have it, a judgment as to which of several competing theories seems most capable of stimulating further research and creating new knowledge?

If we adopt Hayden White's position—namely, that the validity of the historical text lies in the vision of the world it proposes rather than in any merely factual knowledge or understanding it provides—then we also commit ourselves to the view that the two realms of fact and value, knowledge and decision or will, are distinct and uncommunicating. If, on the other hand, we adopt the second position—that there are rational grounds for preferring one theory to another, however difficult it may be to describe precisely how the process of persuasion works—we are committed to no such assumption. To the contrary, no radical distinction, no breach of continuity is assumed between knowledge and decision. Thus, according to Hans Albert, the process of cognition in science is "shot through with regulations, valuations, and decisions [. . . .] We choose our problems, evaluate solutions to them, and decide to prefer one of the available solutions to others." In this way, "ultimately decisions lie 'behind' all knowledge." Similarly, Lakatos acknowledges that in adopting a theory or research program, a scientist *decides* to accepts its hard core—that part of it which Lakatos describes as the "negative heuristic" because it is the part that *may* not be modified or altered. But if "knowledge as a whole seems to slip" as a result of this insight into the role played by decision, and "its objectivity to become questionable," so, conversely, does "the hitherto unanalyzed and unexplained equation of decision and arbitrariness—the thesis of the fundamental irrationality of all decision."[115] It becomes possible to consider whether value judgments themselves need be located entirely outside the realm of rationality. "It is true that one cannot without more ado deduce a value statement from a factual statement," Hans Albert—whose argument I have been following closely here—concedes. "But particular value judgments can certainly turn out to be incompatible with previously held value convictions in the light of a revised factual conviction."[116]

Mommsen's idea of how historical knowledge "progresses" and one account supplants another may not, in sum, be entirely adequate. Nevertheless, I have cited it as symptomatic of the professional historian's general notion of the enterprise he considers himself engaged in. And it is not any more inadequate than the notions some natural scientists entertain about the way their knowledge changes. What is worth noting is that it is the rationality of the historian's activity—"the constant 'feedback' process that subjects to examination the underlying hypotheses, theorems, and explanatory models, and finally results in a re-examination of the ideological or social premises underlying these concepts"[117]—that for Mommsen defines historical thought and gives it its social significance and value. History as a discipline, he argues, subjects prevalent historical assumptions to rational analysis, and thereby tests the validity of the understanding that social groups have of themselves. Most importantly, this testing process is not dependent on the historian's own ideology, by which Mommsen presumably means that though ideology may in part determine the choice of the question to be analyzed, it does not determine the process or the outcome of the

---

[115]Hans Albert, *Treatise of Critical Reason*, 77.
[116]*Treatise of Critical Reason*, 100.
[117]Mommsen, 34.

analysis. Mommsen emphasizes that this clarifying function does not make professional historiography a neo-conservative instrument for filtering out extreme positions. "Scientific neutrality," in other words, "does not reduce the whole range of possible views to some golden mean corresponding to liberal-democratic politics. Its aim is simply to work toward a clearer rational understanding of the positions themselves" and ultimately "to show opposing social and political groups the way toward pragmatic compromises and peaceful resolutions of conflict." History does this, according to Mommsen, "by creating better conditions for maximum communication across ideological lines or, indeed, by first creating conditions of communication as such."[118] In this connection it is worth recalling that the idea of mediating conflict through argument and discussion presided over the very beginnings of modern historiography. I quote from a recent study of Bayle:

In Bayle and his contemporaries we can observe the emergence of a project that is the exact opposite of the one motivating contemporary thought. Our aim is to expose the manifestation of power and the confrontation of competing forces behind the notions of law, meaning, and truth. At that time, the object was to disengage knowledge from power struggles and to disarm the violence of confrontation by establishing a truth of fact that would dissipate the aggressiveness of the pronouncements brandished by the parties in conflict. For as soon as truth ceases to impose itself as an atemporal absolute, but is made subject to a series of mediations, all of which have to be scrupulously examined, there is a displacement of the point at which it intervenes. More precisely, two types of truth are opposed: a truth of faith that demands total spiritual consent and adherence and may therefore exercise physical constraint on the recalcitrant; and a truth of fact, which, by situating the truth of faith inside history [i.e. for Bayle, inside an historical *body* of texts] and considering it as a simple question of fact, deprives it of its relation to the absolute and thereby of all right to exercise physical constraint."[119]

As a modern professional discipline, history, it seems to me, is deeply engaged in that process of rationalization and "disenchantment" that Weber considered "the fate of our times." As such it follows rules and procedures which appear to be distinct from those of artistic creation. In many respects, it seems, as I suggested earlier, that history's deepest affinity is to the law. In a review of two recent historical films—Edgar Reitz's *Heimat* and Claude Lanzmann's *Shoah*—in *the New York Review of Books* not long ago (19 December 1985), Timothy Garton Ash underlined the selectiveness of the memory that presided over the making of both works. In Reitz's case, the selective memory of his characters is integrated into his film as its very subject. Lanzmann, on the other hand, presents his film more explicitly as a work of history and thus raises in a more acute form the question of the relation between history

---

[118]Mommsen, p. 35. see also Jörn Rüsen, "Historische Erinnerung und menschliche Identität," p. 400, on history as "das friedliche Mittel des vernünftigen Argumentierens."
[119]Weibel, *Le Savoir et le corps*, 21.

and art. He acknowledges, for instance, that he threw out some of the hundreds of interviews he taped because the interviewees were "weak" as characters; and on one occasion he described his film tantalizingly as a "fiction of reality" [. . .] "made out of my own obsessions." In both films, but especially in Lanzmann's therefore, Ash discovers that the relation between what he calls "the artistic truth" and "the historian's truth," between "artistic completeness" and "historical completeness," between history and memory, is the central problem. The importance, and the poignancy, of the distinction are well conveyed in the comment on which Ash closes his review. It expresses more effectively than I could myself the reason why I believe it is important to emphasize the rationality of the historical enterprise and the commensurability of historical narratives, their vulnerability to criticism and review.

The one conclusion to which [both films] lead me is: Thank God for historians! Only the professional historians, with their tested methods of research, their explicit principles of selection and use of evidence, only they can give us the weapons with which we may begin to look the thing in the face. Only the historians give us the standards by which we can judge and "place" *Heimat* or *Shoah*. Not that any one historian is necessarily more impartial than any one film director. But (at least in a free society) the terms of the historian's trade make them responsible and open to mutual attack, like politicians in a democracy, whereas the film director is always, by the very nature of his medium, a great dictator. So the historians are our protectors. They protect us against forgetting—that is a truism. But they also protect us against memory.[120]

I am aware that the position I am moving toward may have political and ideological implications. I cannot be blind to the analogy between my idea of an intellectual system which includes mechanisms of adaptation and self-correction and a familiar liberal vision of society, politics, and economics. By choosing to consider historiography as an evolving system of argument, exchange, criticism, and self-criticism, rather than as a collection of colliding, uncommunicating, and incommensurable world-views, I am no doubt signaling not only a belief in underlying continuities, but an ethical preference for evolution and reform rather than revolution, for dialogue and compromise rather than violence. I must simply accept this, constantly examine and re-examine my motives, and consider with as open a mind as possible all objections and criticisms.

---

[120]I should add that, though I agree with the substance of Ash's reflections on *Heimat* and *Shoah*, I do not altogether agree with his reading of *Shoah*. It seems to me that Lanzmann was trying to avoid the unifying and totalizing interpretations that both the artistic shaping of memory and the explanatory discourse of history am to provide. *Shoah* is an oppressive film precisely because it refuses us the reassuring patterns of art and science alike, and Lanzmann's entire enterprise—to represent what is beyond representation and understanding—strikes me as paradoxical, courageous, and in the circumstances entirely appropriate.